CAR-T Manufacturing

This edited volume describes innovations in all steps of the CAR-T manufacturing chain. These vital descriptions will help practitioners to overcome current challenges during the process and vastly reduce costs and enable timely and accessible administration of CAR-T therapy to patients.

The book provides readers with information on key fundamental concepts of CAR-T manufacturing in areas such as cell selection, cell activation, cell transfection, cell expansion, genetic engineering, and quality control. In each chapter, a particular technological field in the CAR-T manufacturing chain is discussed. Each chapter will include an introduction to the importance of a particular technology for cell manufacturing, comparisons of state-of-the-art methods, and discussions on respective emerging innovations. This exposes readers to a high-level view of the entire process while diving into details for each specific process step. Readers will be able to apply their knowledge to make changes at each step of the CAR-T manufacturing process to reduce the existing high costs and long production times, so that cancer patients globally can benefit from CAR-T therapy.

This book is an invaluable resource for practitioners in CAR-T manufacturing who aim to improve their quality and efficiency while reducing time and costs. It is also useful for advanced undergraduate and graduate students who wish to gain a strong foundation for continuing research in the field or interacting with practitioners.

Andy Kah Ping Tay received his Ph.D. from the University of California, Los Angeles (2017) and subsequently completed his postdoctoral training at Stanford University (2019). He then went to Imperial College London as an 1851 Royal Commission Brunel Research Fellow (2020). He currently holds the Presidential Young Professorship in the Department of Biomedical Engineering at the National University of Singapore.

CAR-T Manufacturing
Technologies and Innovations

Edited by Andy Kah Ping Tay

CRC Press
Taylor & Francis Group
Boca Raton London New York

CRC Press is an imprint of the
Taylor & Francis Group, an **informa** business

Designed cover image: T-Cells Work to Fight Cancer, Immunotherapy, CAR-T cell therapy, 3d rendering; Shutterstock

First edition published 2025
by CRC Press
2385 NW Executive Center Drive, Suite 320, Boca Raton FL 33431

and by CRC Press
4 Park Square, Milton Park, Abingdon, Oxon, OX14 4RN

CRC Press is an imprint of Taylor & Francis Group, LLC

ISBN: 978-1-032-66076-9 (hbk)
ISBN: 978-1-032-66073-8 (pbk)
ISBN: 978-1-032-66075-2 (ebk)

DOI: 10.1201/9781032660752

Typeset in Sabon
by SPi Technologies India Pvt Ltd (Straive)

Contents

Contributors

Arun R. K. Kumar

Arun is a Ph.D. candidate at the National University of Singapore supported by the iHealthtech Research Scholarship. Arun's research interests include fabrication of high-aspect-ratio nanostructures and non-viral transfection of primary immune cells. He received his B. Tech in Biotechnology degree from Vellore Institute of Technology, India.

Kenneth Jun Lim Goh

Kenneth Goh is an undergraduate student at the National University of Singapore. He is currently studying for a Bachelor of Engineering (Biomedical Engineering), graduating in 2025.

Jessalyn Low

Jessalyn Low is a Ph.D. candidate at the National University of Singapore. She received her B.Eng. degree in Materials Engineering from the Nanyang Technological University in 2020.

Yikai Luo

Yikai Luo is a postgraduate student in the Department of Biomedical Engineering at the National University of Singapore. His research interests focus on developing biomedical devices that turn concepts to reality. His current project under Andy Kah Ping Tay's team is to develop an automated bioreactor for chimeric antigen receptor T cell therapy.

Yu Yang Ng

Yu Yang received his Ph.D. from the National University of Singapore (2019), with focus on immune cell engineering for immunotherapy. His current research focus is investigating age-associated changes to the immune system.

Mohamed Bilal S/O Saibudeen

Mohamed Bilal is doing his bachelor's degree in Biomedical Engineering specializing in Tissue Engineering, with a minor in Biophysics at the National University of Singapore. He earned a silver medal for his Diploma in Bioengineering from Singapore Polytechnic in 2022, graduating with merit.

Andy Kah Ping Tay

Andy Kah Ping Tay received his Ph.D. from the University of California, Los Angeles (2017), and subsequently completed his postdoctoral training at Stanford University (2019). He then went to Imperial College London as an 1851 Royal Commission Brunel Research Fellow (2020). He currently holds the Presidential Young Professorship in the Department of Biomedical Engineering at the National University of Singapore. With his team, he is developing technologies and material to engineer the immune system for cancer treatment.

Jerome Adriel Tjiptadi

Jerome Adriel Tjiptadi is a biomedical engineering doctoral student at the National University of Singapore, working under the supervision of Dr. Andy Kah Ping Tay. He received his bachelor's degree in Materials Engineering from Nanyang Technological University in 2021. His research interests include tissue engineering and bioinspired materials for cancer immunotherapy.

1   Immune cell selection and isolation

Jessalyn Low and Andy Kah Ping Tay

Introduction

The field of cancer immunotherapy has been expanding rapidly in the past decade, with CAR-T cell therapy emerging as a groundbreaking approach for the treatment of various malignancies, particularly in the treatment of B cell leukemia and lymphoma. Despite its remarkable successes, there remain major limitations and barriers to the use of CAR-T cell therapy, for instance, challenges in its safety and efficacy, as well as other manufacturing limitations. The challenges to more widespread adoption of CAR-T cells in the clinic have spurred the investigation of alternative immune cell sources for CAR therapy.

In recent years, cell sources like natural killer (NK) cells, macrophages, and unconventional T cells, including γδ T cells, MAIT cells, and natural killer T (NKT) cells, have emerged as promising candidates as novel cell sources for CAR therapy. These cell sources offer unique advantages, thereby providing promising avenues to overcome the limitations of CAR-T cell therapy.

Here, we discuss the sources and characteristics of these alternative cell sources to provide insights into their applications for CAR immune cell therapy.

CAR-T cell therapy

The field of CAR-T cell therapy has experienced remarkable growth and advancement in recent years, with the first CAR-T cell approved by the Food and Drug Administration (FDA) in 2017. To date, there are six CAR-T cell therapies approved by the FDA – Kymriah, Yescarta, Tecartus, Breyanzi, Abecma, and Carvykti. All six of these are autologous CAR-T cell products targeting hematological malignancies, primarily acute lymphoblastic leukemia (ALL), non-Hodgkin lymphoma (NHL), and multiple myeloma.

There are also more than 1000 ongoing clinical trials targeting hematological malignancies like ALL, NHL, multiple myeloma, chronic lymphocytic leukemia (CLL), and AML, as well as solid tumors such as in pancreatic, lung, and breast cancer.

Sources of CAR-T cells

CAR-T cell manufacturing typically starts from the collection of peripheral blood mononuclear cells (PBMCs) from the individual, typically via leukapheresis. T cells are then purified from the PBMCs. CAR-T cells are generated from CD3+ T cells, consisting of both CD4+ helper T cells and CD8+ cytotoxic T cells at variable ratios.

DOI: 10.1201/9781032660752-1

CAR-T cell products generated with defined compositions can provide uniform potency compared with products derived from unselected T cells that differ in phenotypic composition.[1] At a 1:1 CD4:CD8 ratio, CAR-T cells were found to exhibit improved cell expansion, persistence, and toxicity, and to be associated with a higher remission rate.[2] It has also been suggested that coculturing both the CD4+ and CD8+ cells together could help maintain robust CD8+ CAR-T cell function, compared to when they are cultured separately.[3]

Because CD4+ T cells and CD8+ T cells exert different functions, studies have also attempted to determine the optimal CD4:CD8 ratio of CAR-T cell products that will produce synergistic antitumor effects and enhance clinical outcomes. CD8+ T cells are known to be important as they exert their cytotoxicity against the target tumor cells through a perforin-dependent manner, while CD4+ T cells promote CD8+ T cell function, and can also elicit direct antitumor cytotoxicity, although the mechanisms have not been well studied. It has been demonstrated that the CD4+:CD8+ ratio is a prognostic factor for efficacy and toxicity, where a higher percentage of CAR+ CD8+ cells was found to be a significant predictor for response at three- and six months following CAR-T treatment, suggesting the importance of CD8-mediated cytotoxic mechanisms of CAR-T cells in achieving therapeutic efficacy.[4]

Taking the example of FDA-approved lisocabtagene maraleucel (Breyanzi), the FDA has stipulated a recommended 1:1 ratio of CD4 and CD8 components. CD4+ T cells and CD8+ T cells are immunomagnetically selected from leukapheresis products, with the purified CD4+ and CD8+ T cells separately activated, transduced, and expanded. Each component is supplied separately, with the CD4 component being supplied immediately after administration of the CD8 component.[5]

Autologous and allogenic CAR-T therapy

Broadly, depending on the source of T cells, CAR-T cell therapy can be categorized into autologous and allogenic. Autologous CAR-T therapy uses cells derived from the patient's own T cells, while allogenic CAR-T therapy uses T cells derived from healthy donors rather than the patient's own cells.

Traditionally, CAR-T cell therapy has focused on autologous cell-based therapy, due to its lowered risk of immune rejection and graft-versus-host disease (GvHD) as the CAR-T cells are derived from the patients. However, autologous CAR-T therapy is associated with various challenges like complex manufacturing processes and potential delays in treatment availability. Lengthened vein-to-vein time is undesirable, particularly for patients with rapidly progressing disease. Autologous T cells may also not be effective in some patients owing to T cell dysfunction like T cell exhaustion and senescence, which are more often observed in cancer patients due to the immune effects of cancer or exposure to chemotherapeutic agents.[6] Poor T cell quality could furthermore give rise to manufacturing failures, for instance, due to poor in vitro expansion, resulting in insufficient CAR-T cell products. Taken together, delays in treatment affect not only access to CAR-T treatments but also treatment efficacy. In a study examining the value of reducing wait times for CAR-T cell treatment, it was shown that this was associated with a 14% increase in treatment efficacy.[7]

To make treatment more affordable and readily available, there has been an increased interest in the use of allogenic CAR-T, also known as 'off-the-shelf' CAR-T cells, where CAR-T cells are derived from healthy donors. From the manufacturing perspective, allogenic CAR-T cell therapy allows for scaling up of manufacturing processes while decreasing costs. Besides, from the treatment perspective, allogeneic CAR-T cells produce batches of cryopreserved T cells, making treatments immediately readily available for patients and reducing wait times. This also allows for redosing, if necessary.

Challenges of CAR-T therapy

While numerous promising successful cases of CAR-T cell therapy in treating cancer patients have been reported, CAR-T therapy still faces many challenges and barriers, both in terms of safety and efficacy.

CAR-T cell therapy can induce CAR-T-associated toxicities, with high rates of toxicities and fatalities hindering its adoption as the first line of treatment. The most common toxicity following CAR-T infusion is cytokine release syndrome (CRS), which has been reported to be as high as 100% in some clinical trials.[8] CRS is a cytokine-mediated systemic inflammatory response due to immune activation, which results in elevated serum cytokine levels that can be life-threatening. Immune effector cell-associated neurotoxicity syndrome (ICANS) is also a primary toxicity associated with CAR-T that often accompanies CRS but has also been reported to occur independently. Typically, tocilizumab, an interleukin-6 (IL-6) antagonist, is used as first-line treatment for severe or life-threatening CRS, while corticosteroids are used as first-line therapy for isolated ICANS and also for life-threatening or tocilizumab-resistant CRS.[9,10]

Another safety concern is on-target, off-tumor toxicities, where CAR-T cells recognize and attack healthy tissues expressing the target antigen, which could potentially lead to unintended tissue damage and adverse effects.

Moreover, allogenic CAR-T cell therapy also introduces additional challenges, in particular alloimmune rejection and GvHD. The T cell receptor (TCR) from donor CAR-T cells is able to recognize foreign human leukocyte antigen (HLA) molecules with peptides restrictively, hence inducing acute or chronic GvHD. Host T cells may also attack allogeneic CAR-T cells in an HLA-restricted way, thus triggering host rejection.

Besides issues pertaining to the safety of CAR-T cell therapy, challenges to its efficacy are present. Antigen escape poses a significant challenge, particularly in the context of hematological malignancies. Tumor cells may downregulate or lose the expression of the target antigen, hence restricting the efficacy of CAR-T therapy. For instance, 7%–25% of patients in CD19 CAR-trials for ALL show CD19⁻ relapse.[11] Prolonging CAR-T cell persistence is also a key challenge to tackle. Short-term CAR-T cell persistence leads to a weak response to CAR-T cell therapy and is associated with poor efficacy.

The use of CAR-T cells for the treatment of solid tumors also poses a unique set of efficacy challenges, resulting in fewer successes demonstrated for solid tumors compared to liquid tumors. Notably, the immunosuppressive tumor microenvironment (TME) creates a hostile environment for CAR-T cells, causing T cell dysfunction and limiting infiltration into the solid tumor. In addition, the ability of CAR-T cell trafficking into solid tumors is limited by barriers such as poor vascularization and physical barriers like cancer-associated fibroblasts (CAFs) within the tumor tissue.

Novel cell sources for CAR therapy

CAR-T cell therapy has demonstrated remarkable efficacy in treating certain types of cancer, but the use of T cells has also been associated with various challenges. Overcoming these challenges is critical for the success of CAR therapy. To this end, various strategies

Table 1.1 Comparison of CAR-T cells, CAR-NK cells, and CAR-macrophages

	CAR-T	*CAR-NK*	*CAR-Macrophages*
Source	Peripheral blood, bone marrow, umbilical cord blood	Peripheral blood, iPSCs, HSPCs, umbilical cord blood, NK-92 cell line	Peripheral blood monocytes, iPSCs, HPSCS, bone marrow–derived monocytes
Mechanism of action	CAR-dependent cytotoxicity	CAR-dependent cytotoxicity, NK-mediated cytotoxicity	CAR-dependent phagocytosis, macrophage-mediated remodeling of TME
Tumor infiltration ability	Good for hematological malignancies, poor for solid tumors	Moderate	Moderate to good, able to infiltrate TME
Persistence	Long-term; potential for memory	Short to medium-term; longer with cytokine pre-exposure, may require re-dosing	Short to medium-term; may require re-dosing
Cytotoxicity	High	Moderate to high	Moderate
Manufacturing complexity	High, requires significant in vitro expansion to achieve therapeutic doses; typically autologous which requires personalized processes	Moderate, but still requires development of expansion protocols	High, requires differentiation from monocytes or iPSCs, more challenging for large-scale production
Off-the-shelf potential	Limited, challenges of GVHD and immune rejection	High, but with challenges in cryopreservation	High
Safety profile (CRS, neurotoxicity)	High risk	Lower risk	Lower risk

have been employed, for instance, by altering the CAR design or by engineering the CAR-T cells. One other key approach is to use other immune effector cell types for CAR therapy. Different immune cells have unique properties (Table 1.1) and also work differently to kill the target tumor cells (Figure 1.1), which could be valuable to overcoming the limitations brought about by T cells. Numerous startups and biotechnology companies have also been set up to develop CAR products from alternative cell sources (Table 1.2). Here, we discuss alternative cell sources for CAR therapy.

CAR-NK

CAR-NK therapy has gained much interest in recent years as a promising alternative to CAR-T therapy. CAR-NK cells are derived from NK cells, which are innate immune cells found in peripheral circulation. NK cells possess inherent cytotoxic activity and are functionally similar to cytotoxic T cells, but without a somatically rearranged and antigen-specific TCR.[13] NK cells, hence, exert their cytotoxic activity through a balance between activating and inhibitory signals, without any prior sensitization or antigen recognition.

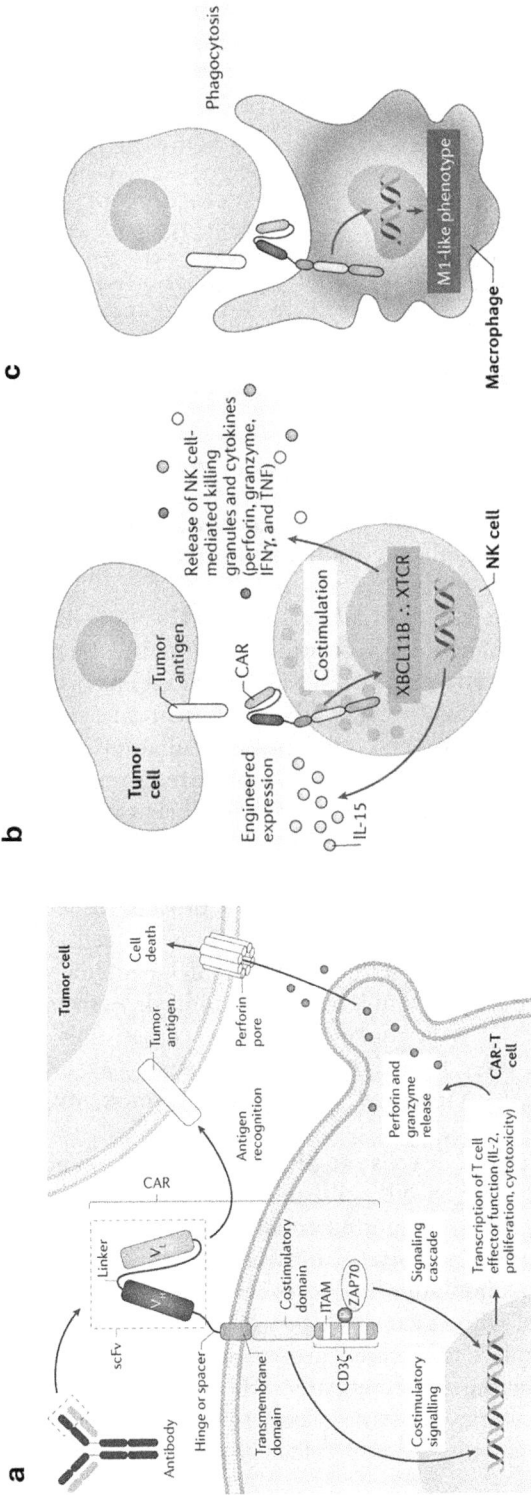

Figure 1.1 Mechanism of action of CAR-T cells, CAR-NK cells, and CAR-macrophages. NK cells and macrophages have emerged as novel cell sources to overcome limitations of CAR-T cells. These cell types have different mechanisms of action compared to T cells, conventionally used for CAR-T cell therapy. (a) CAR-T cells are activated upon scFv recognition of the tumor antigen, causing clustering and immobilization of CAR-molecules. A signaling cascade is initiated, resulting in T cell effector response and cytotoxic function through secretion of granzyme and perforin, causing cell death of tumor cells. (b) CAR-NK cells kill tumor cells by releasing cytotoxic granules like perforin and granzymes and performing antibody-dependent cellular cytotoxicity (ADCC) via Fc receptors. Engineering CAR-NK cells to express IL-15 further enhances antitumor response. (c) CAR-macrophages induce phagocytosis of antigen-positive tumor cells. CAR-macrophages can also present tumor antigens to T cells, further promoting adaptive immunity against the tumor. Figure adapted from Larson and Maus (2021)[12] with permission.

Table 1.2 Startups and biotechnology companies in the area of emerging CAR-immune cells

Name	Focus	Key products and target
Carisma Therapeutics	CAR-macrophages, CAR-monocytes	CT – 0525 (HER2 CAR-monocyte)
Century Therapeutics	iNK	CNTY-101 (CD19 CAR-iNK)
Nkarta	CAR-NK	NKX019 (CD19 CAR-NK)
Cytovia Therapeutics	iNK, CAR-iNK	CYT-503 (GPC3 CAR-iNK)
Artiva Biotherapeutics	NK, CAR-NK	AB-201 (HER2 CAR-NK)
Fate Therapeutics	CAR-T, CAR-NK	FT522 (CD19, 4-1BB CAR-NK)
Senti Bio	CAR-NK	SENTI-202 (CD33 and/or FLT3 CAR-NK); SENTI-301A (GPC3 CAR-NK)
Appia Bio	CAR-NKT, CAR-γδT	API-192 (CD19/20 CAR-NKT)

Sources of CAR-NK cells

Many sources of NK cells for CAR-NK cell production have been studied, including NK-92 cell line, peripheral blood, umbilical cord blood, and induced pluripotent stem cells (iPSCs).

Due to the unlimited proliferative ability of NK-92 cells, CAR-NK cells from NK-92 can be generated at a large scale, unlike the use of primary NK cells, which carries challenges in isolation and expansion ex vivo. However, NK-92 is a cell line originating from a patient with non-Hodgkin's lymphoma, bringing about inherent safety risks of potential tumorgenicity. NK-92 cells, therefore, require prior irradiation before infusion, which results in a loss of in vivo proliferative abilities. Compared to using primary NK cells, in vivo persistence and overall therapeutic potential are likely to be lower.

Therefore, while NK-92 cells have been used in multiple clinical trials, more studies have moved toward the use of primary NK cells due to their superior functional abilities, despite facing critical challenges in the ex vivo expansion process to achieve clinical doses. Peripheral blood is one of the most common sources of NK cells for CAR-NK therapy, where NK cells form typically 5%–15% of peripheral T lymphocytes. NK cells can be obtained from PBMCs, and then stimulated and expanded. Activated CAR-NK cells derived from PBMCs express a wide range of active receptors, thus allowing for expansion in vivo. Over 90% of these PBMC-derived NK cells are CD56^{dim}CD16$^+$ NK cells, typically exhibiting properties of increased maturity and cytotoxicity, yet reduced proliferative capacity.

Similar to PBMC-derived NK cells, NK cells can also be obtained from umbilical cord blood (CB). Compared to PBMC-derived NK cells, these cells are typically less mature with lower cytotoxicity, yet, they have a greater potential for ex vivo expansion. In addition, CB-derived cells have shown the potential to differentiate into functionally mature effector cells following ex vivo co-stimulation with cytokines.[14]

To tackle challenges associated with expansion, the use of iPSC-derived NK cells has of late gained much interest. Owing to their clonal growth and high expansion potential, large numbers of mature homogenously differentiated NK cells can be generated, thus allowing for large-scale manufacturing operations. Compared to PBMC- and CB-derived NK cells, iPSC-derived NK cells can also be more efficiently engineered for stable CAR expression.

Advantages and challenges

One key advantage of using CAR-NK cells is that unlike T cells, NK cells do not cause GvHD in the allogeneic setting. NK cells recognize their targets in an HLA-unrestricted manner; therefore, CAR-NK cells do not require HLA compatibility. As such, CAR-NK cells can be generated from a wide range of autologous and allogenic sources. This opens the potential for generating off-the-shelf allogenic therapies that can be mass-produced and made readily available.

In addition, compared to CAR-T cells, CAR-NK cells have been shown to display superior safety where clinical trials demonstrate that they are less associated with adverse effects such as CRS and neurotoxicity, attributed to the difference in cytokines produced. Due to the lower lifespan of CAR-NK in circulation, the risk of on-target, off-tumor toxicity is also lowered.

Furthermore, unlike CAR-T cells, CAR-NK cells can exert targeted cytotoxicity both in a CAR-dependent and CAR-independent manner. While the CAR constructs enable more rapid targeting of tumor cells, NK cell receptors can still activate the cytotoxic activity of CAR-NK cells independently of the CAR construct. As such, CAR-NK cells could exert more flexibility in the clearance of a heterogeneous tumor where some tumor cells do not express the target antigen of the CAR construct through CAR-independent mechanisms, serving as a potential approach for tackling tumor escape. The CAR-independence mechanisms can also be leveraged when paired with careful selection of target antigens to minimize off-target effects. This is, for instance, by engineering a non-signaling CAR construct that instead promotes NK cell homing and adhesion to targets, which then allows the NK cells to exert their natural cytotoxicity abilities.

It should be noted that CAR-NK cells, however, have limitations which should be tackled to further the use of CAR-NK in cancer immunotherapy. For instance, CAR-NK cells have a short lifespan due to their lower in vivo persistence, which limits therapeutic efficacy, albeit benefiting safety. To improve persistence, strategies such as the use of cytokines like interleukin-12 (IL-2) and interleukin-15 (IL-15) are being studied for signaling through the common gamma (γ_c) chain and promote cell proliferation.[15]

In addition, various steps of the cell manufacturing process have to be optimized to tailor to the inherent challenges brought about by NK cells. For instance, ex vivo expansion and activation of NK cells are especially critical to generate clinical doses. NK cells are also sensitive to the freeze-thaw process, which compromises their viability and cytotoxicity post-thawing. More studies into optimizing these processes are thus integral, for instance, through the use of feeder cells and cytokines.

Clinical progress

CAR-NK cell therapy has shown promising results in both the treatment of hematological malignancies and solid tumors. In a first-in-man clinical trial of CAR-NK-92 cells in patients with relapsed and refractory AML, it was found that at doses up to five billion cells per patient, no significant adverse effects were observed. Additionally, one of the most significant breakthroughs in the field of CAR-NK was the results of a phase 1 and 2 study of NK cells of the successful administration of HLA-mismatched anti-CD19 CAR-NK cells derived from cord blood to 11 patients with relapsed or refractory CD19-positive cancers. Eight out of 11 treated patients had a response and without major toxic effects, with 7 of them having a complete remission.[16]

There have also been many ongoing clinical trials using CAR-NK cells for the treatment of solid tumors, such as glioblastoma, breast cancer, lung cancer, gastric cancer, and renal cancer,[17] however, with limited clinical data to date. This includes, for instance, ACE1702 anti-HER2 oNK cells targeting human HER2-expressing solid tumors (NCT04319757) and PD-L1 t-haNK cells targeting locally advanced or metastatic solid tumors (NCT04050709). Some of these clinical trials also explore the combinatory effect with other modalities such as antitumor drugs. This includes, for instance, intracranial injection of NK-92/5.28.z cells in combination with intravenous ezabenlimab in subjects with recurrent HER2-positive glioblastoma (NCT03383978), and irradiated PD-L1 CAR-NK cells combined with pembrolizumab and N-803 for subjects with recurrent or metastatic gastric or head and neck cancer (NCT04847466).

CAR-macrophages

Macrophages are innate immune cells and are plastic in nature, exhibiting diverse functional phenotypes dependent on their environmental cues. Traditionally, macrophages have two phenotypes – classically activated M1-like macrophages that are involved in pro-inflammatory responses and alternatively activated M2-like macrophages that are involved in anti-inflammatory responses. Tumor-associated macrophages (TAMs) form a distinct subpopulation of macrophages residing in the TME and are also the most abundant innate immune cells in the TME. By exploiting their infiltration capabilities into solid tumors, CAR-Ms have emerged as potential candidates especially for solid malignancies.

Sources of CAR-macrophages

Cell lines like THP-1, a human leukemia monocytic cell line, can be used as sources of macrophages. THP-1 monocytes can be differentiated into M1-like macrophages by phorbol myristate acetate (PMA) stimulation and treated with lipopolysaccharides (LPS) and IFN-γ.

For autologous cell therapies, peripheral blood monocytes can be isolated from peripheral blood via leukapheresis as primary macrophages are not abundantly found in blood. Isolated monocytes can then be differentiated to macrophages using cytokines like macrophage colony-stimulating factor (M-CSF) or granulocyte-macrophage colony-stimulating factor (GM-CSF). However, large quantities of monocytes would be required, which may potentially be a challenge especially in cancer patients with reduced peripheral monocytes.

Similar to CAR-NK, iPSCs can also be used as a source for macrophages, where iPSCs are collected from patients and then differentiated. In addition, another source is hematopoietic stem and progenitor cells (HSPCs) isolated from bone marrow or cord blood. This is especially useful for allogenic cell therapies and for large-scale manufacturing to generate high amounts of functional CAR-Ms, which can address issues concerning limited cell sources.

Advantages and challenges

Macrophages possess a remarkable capability to infiltrate solid tumor tissue and interact with the diverse cellular components within the complex tumor TME. Due to the immunosuppressive nature of the TME, CAR-T cell trafficking and infiltration into the solid tumor are severely hampered. Suppressive immune cells like regulatory T cells (T_{reg}) and myeloid-derived suppressor cells hamper CAR-T cell responses. In addition, the structure of the solid tumor stroma poses a physical barrier to CAR-T cell infiltration. This is unlike macrophages that can efficiently infiltrate the TME and in fact are the most abundant innate immune cells recruited to the tumor site as they play critical roles like stimulating angiogenesis and enhancing tumor invasion. Their adeptness in penetrating and navigating the immunosuppressive TME therefore makes CAR-M promising candidates for cancer immunotherapy particularly for solid tumors. By engineering CAR-Ms to enhance effector functions like targeted phagocytosis, antigen presentation, and cytokine secretion, CAR-Ms are versatile platforms for CAR therapy.

Although CAR-Ms hold much potential for cancer immunotherapy, some limitations need to be overcome for it to be realized. Similar to CAR T therapy, the high heterogeneity of tumor cells may mean that target antigen expression may be insufficient. The majority of targeted tumor antigens are also found on healthy cells, which could lead to off-target toxicities. In addition, most exogenous macrophages remain in the liver, which could lead to other toxicities.

Furthermore, while macrophages possess the capability for tumor infiltration, macrophages have high plasticity to adapt to their environmental stimuli, where cytokines within the TME may repolarize antitumor M1 macrophages to pro-tumor M2 macrophages.

Achieving clinical doses may also be challenging due to the limited proliferative capacity of macrophages in vitro and in vivo particularly when relying on primary peripheral blood monocytes as a source for macrophage. This, however, may be overcome with approaches that use iPSCs or HSPCs to generate sufficient amounts of CAR-Ms.

Clinical progress

To date, there have been few clinical trials of CAR-Ms. Most notable is the first-in-human, open-label study of CAR-M in HER2-overexpressing solid tumors, CT-0508, developed by Carisma Therapeutics Inc. (NCT04660929). In a 2022 update of the phase 1 clinical trial, the data showed that CT-0508 remodeled and activated the TME and initiated antitumor T cell immunity, while a 2023 update showed that CT-0508 was successfully manufactured for patients and that the administration was well tolerated after infusion with no dose-limiting toxicities reported to date. Phase 1 study of CT-0508 in combination with KEYTRUDA® (pembrolizumab) in patients with HER2 overexpressing solid tumors has also been initiated to tackle the increased T cell exhaustion observed.

Unconventional T cells for CAR therapy

While conventional CD4+ and CD8+ T cells have been extensively studied, emerging evidence suggests that unconventional T cell subsets also play significant roles, which collectively constitute approximately 10% of circulating T cells. Notable subsets include NKT cells, mucosal-associated invariant T (MAIT) cells, and γδ T cells. These unconventional T cell subsets differ in their antigen recognition mechanisms, TCR composition, functional properties, and tissue distribution, reflecting their distinct roles in immune surveillance and host defense. With their unique properties, they have also emerged as potential candidates for cancer immunotherapy.

CAR-γδT

γδ T cells are a subset of T cells which express a TCR γ-chain in combination with a TCR δ-chain, unlike conventional αβ T cells where TCR comprises a heterodimer of α and β chains. γδ T cells are a highly heterogeneous group of cells, exhibiting different properties based on their γ and δ chains. γδ T cells originate from the thymus, where they undergo development and maturation. However, unlike αβ T cells which primarily traffic to secondary lymphoid organs like lymph nodes and the spleen for clonal expansion and activation upon encountering antigens presented by antigen-presenting cells (APCs), many γδ T cells migrate directly to peripheral tissues or mucosal sites where they encounter antigens. As such, γδ T cells show reduced clonal expansion and TCR diversity than αβ T cells. While γδ T cells account for 0.5%–5% of peripheral blood T cells, they form a major lymphocyte population on mucosal and epithelial surfaces.

There has been interest in using γδ T cells for CAR-γδ T cell therapy due to their numerous advantages over conventional αβ T cells. First, γδ T cells exhibit broad antigen recognition. Unlike αβ T cells which recognize peptide antigens presented by major histocompatibility complex (MHC) molecules, γδ T cells do not rely on MHC presentation for antigen recognition and can recognize a wide range of antigens, including non-peptide antigens, for instance, stress-induced molecules or tumor-associated antigens, allowing for broader tumor reactivity.

Next, γδ T cells are able to directly engage with tumor cells through their innate-like characteristics and are also capable of stimulating other immune cells like dendritic cells (DCs) and NK cells via cytokine secretion to induce a cascade of immune responses to amplify antitumor responses and enhance tumor toxicity.

γδ T cells, in particular the Vδ1 subset, possess tissue-homing ability to various tissues. γδ T cells are enriched in epithelial tissues and mucosal sites. By harnessing these tissue-homing properties, CAR-γδ T cells demonstrate promising potential in infiltrating these solid tumor sites and improving their capacity to target tumor cells. In addition, γδ T cells can recognize their target cells in an MHC-independent manner, thus not causing GvHD, making them a potential source for allogenic CAR-γδ T cell therapy.

One challenge of CAR-γδ T cells is that while γδ T cells are enriched in peripheral tissues, their abundance in peripheral blood is low and furthermore may vary among individuals, which could impact their availability for CAR-γδ T cell therapy. In addition, γδ T cells exhibit less clonal expansion and TCR diversity compared to αβ T cells, which may hinder their proliferative capacity and range of antigen recognition.

Several clinical trials using CAR-γδ T cells are currently underway to study their potential for cancer immunotherapy. For instance, CytoMed Therapeutics Limited started a phase 1 clinical trial in 2023 for allogeneic NKG2DL-targeting CAR-γδ T cells (NCT05302037), to target NKG2D ligands, a form of stress-induced cancer antigens which are commonly present in many cancers.

CAR-MAIT

MAIT cells are an unconventional T cell subset characterized by the expression of a semi-invariant TCR recognizing non-peptide antigens presented by the non-polymorphic MHC-related protein 1 (MR1) molecule. MAIT cells are predominantly found in human tissues and can comprise up to 10% of peripheral T lymphocytes.

One key advantage of CAR-MAITs is their tissue residency, attributed to their high expression of tissue-homing markers. As MAIT cells are enriched in mucosal-associated peripheral tissues such as the gastrointestinal tract, liver, and cervix, where many cancers arise, this tissue residency enhances their capacity to infiltrate and target tumor cells within these sites, potentially enhancing therapeutic efficacy.

Next, MAIT cells also exhibit potent cytotoxic activity upon activation as they can rapidly produce pro-inflammatory cytokines like IFN-γ and TNF-α. This can enhance antitumor immune responses and recruit other immune effector cells to the tumor microenvironment. In addition, MAIT cells are MHC-unrestricted and are thus unlikely to induce GvHD. CAR-MAITs, hence, can be used for allogenic and off-the-shelf cancer immunotherapy.

Some challenges associated with CAR-MAITs include challenges in obtaining clinical doses of CAR-MAITs, due to the low percentage and donor-to-donor variability in MAIT cells in peripheral blood, similar to CAR-γδ T cells. Like T cells, exhausted phenotypes have also been observed in MAIT cells in chronic viral hepatitis, which may reduce the therapeutic efficacy of CAR-MAITs.

Although there have been no clinical trials of CAR-MAITs reported thus far, several studies have demonstrated their potential. In a recent study in 2022, a study showed that CAR-MAITs were found to demonstrate robust cytotoxicity against CD19+ lymphomas and Her2-expressing breast cancer cell lines, while releasing lower levels of inflammatory cytokines, potentially leading to fewer adverse reactions like cytokine release syndrome.[18]

CAR-NKT

NKT cells are a subset of αβ T cells that share surface markers and functional characteristics with both conventional T cells and NK cells. Unlike conventional T cells, NKT cells recognize glycolipid antigens presented by the non-classical MHC class I-like molecule CD1d. NKT cells are broadly classified into two main types based on their TCR diversity and antigen specificities – type I and type II NKT cells. Type I NKT cells, also known as invariant NKT (iNKT) cells, express an invariant TCR composed of a specific α chain paired with a limited repertoire of β chains, while type II NKT cells do not express the invariant TCR and exhibit more heterogeneity in their TCR repertoire. iNKT cells are more well characterized than type II NKT cells and have been shown to play critical roles in immunosurveillance against tumor cells. NKT cells are extremely rare in peripheral blood, with iNKT cells consisting of less than 1% of circulating T cells. NKT cells are also found in the liver and spleen, with iNKT cells also found in the intestine and pulmonary mucosa.

Due to the favorable characteristics of iNKT cells, CAR-NKTs have emerged as potential candidates for CAR-based immunotherapy. Similar to CAR-NK cells, CAR-NKT cells can recognize tumor antigens by both their receptors and CAR structure, therefore being able to exert direct cytotoxicity on the tumor cells. Notably, NKT cells recognize glycolipid antigens presented by CD1d through their TCRs, allowing for recognition of a wide range of tumor-associated antigens owing to the broad expression of CD1d. In addition, expression of NK activating receptors on iNKT cells such as NKG2D provides a secondary method for cell recognition and NK-mediated cytotoxicity independently of CD1d presentation. Their versatility in recognizing tumor antigens increases the potential applicability of CAR-NKT cells across different cancer types.

CAR-NKT cells also possess a natural ability to effectively traffic to the tumor site and do not have the risk of GvHD owing to the lack of MHC engagement, opening up possibilities for CAR-NKT cells to be used in treatment of solid tumors and allogenic cell therapies.

However, similar to the other unconventional T cell types, NKT cells are relatively rare in peripheral blood compared to other cell subsets like NK cells and T cells. Obtaining sufficient NKT cells for CAR engineering may pose challenges for large-scale clinical applications.

One of the landmark studies establishing the potential of CAR-NKT cells was published in 2014, where primary human NKT cells were engineered to express a CAR against the GD2 ganglioside (CAR.GD2), which is highly expressed by neuroblastoma (NB). It was demonstrated that CAR.GD2 NKT cells were effectively localized to the tumor site and had potent antitumor activity, and repeat injections significantly improved the long-term survival of mice with metastatic NB. This has then led to a first-in-human phase 1 clinical trial of GD2-CAR.15 NKTs for relapsed or refractory neuroblastoma. In the updated phase 1 trial interim results, it was shown that the treatment was well tolerated and able to mediate sufficient antitumor activity to produce objective responses in patients with NB.

Future perspectives

The exploration of novel cell sources for CAR therapy offers exciting prospects for advancing the next generation of CAR therapies and enhancing the therapeutic potential of cancer immunotherapy. By harnessing the unique properties of other immune cell types like NK cells, macrophages, and unconventional T cells, efficacy and safety challenges associated with CAR-T cell therapy can be addressed. In particular, these cell types also target key challenges in the development of allogenic off-the-shelf therapies and solid tumor treatments, paving the way for further breakthroughs in cancer treatment. Promising preclinical and clinical results obtained thus far further underscore the potential of these novel cell sources to complement traditional CAR-T cell therapy. It should be noted that while CAR therapy focuses primarily on cancer immunotherapies, recent studies are also exploring the potential for CAR therapy to be used on other diseases such as diabetes, heart failure, and other senescence-associated diseases. Advances in CAR therapy therefore benefit not only cancer immunotherapy but also potentially a wide range of other diseases.

Looking ahead, future research efforts should look into optimizing the engineering and manufacturing processes for these cell types used as CAR therapy platforms. This includes optimizing CAR design to enhance cytotoxicity and persistence, as well as optimization of activation and expansion protocols, which is particularly important for rare cell types. Other CAR cell types like CAR-DCs, CAR-neutrophils, and CAR-B cells should also be further explored. Although current research in these cell sources is still nascent, they may bring about advantages that could make them future avenues for cancer immunotherapy.

Synergistic approaches with different CAR sources should also be explored. Preliminary studies have shown that the administration of CAR-T cells with CAR-NK cells or CAR-Ms is associated with improved clinical outcomes like persistence and tumor-targeting ability. Ultimately, different cell sources bring their own advantages and challenges; as such, combinatorial immunotherapy may potentially be more effective than when relying on CAR cells from a single cell type alone, thus offering opportunities to maximize therapeutic benefits in cancer treatment.

Overall, the exploration of novel cell sources for CAR therapy represents a dynamic and evolving field with great potential to revolutionize cancer treatment. With continued research, these approaches hold the promise of realizing more widespread adoption of CAR therapy for cancer immunotherapy, impacting the lives of patients.

References

1. Sommermeyer, D. et al. Chimeric Antigen Receptor-Modified T Cells Derived from Defined CD8+ and CD4+ Subsets Confer Superior Antitumor Reactivity in Vivo. *Leukemia* **30**, 492–500 (2016).
2. Turtle, C. J. et al. CD19 CAR-T Cells of Defined CD4+:CD8+ Composition in Adult B cell ALL Patients. *J Clin Invest* **126**, 2123–2138 (2016).
3. Lee, S. Y. et al. CD8+ Chimeric Antigen Receptor T Cells Manufactured in Absence of CD4+ Cells Exhibit Hypofunctional Phenotype. *J Immunother Cancer* **11**, e007803 (2023).
4. Galli, E. et al. The CD4/CD8 Ratio of Infused CD19-CAR-T is a Prognostic Factor for Efficacy and Toxicity. *British Journal of Haematology* **203**, 564–570 (2023).
5. Abramson, J. S. et al. Lisocabtagene Maraleucel for Patients with Relapsed or Refractory large B-cell lymphomas (TRANSCEND NHL 001): A Multicentre Seamless Design Study. *The Lancet* **396**, 839–852 (2020).
6. Depil, S., Duchateau, P., Grupp, S. A., Mufti, G. & Poirot, L. 'Off-the-shelf' Allogeneic CAR T Cells: Development and Challenges. *Nat Rev Drug Discov* **19**, 185–199 (2020).
7. Chen, A. J., Zhang, J., Agarwal, A. & Lakdawalla, D. N. Value of Reducing Wait Times for Chimeric Antigen Receptor T-Cell Treatment: Evidence From Randomized Controlled Trial Data on Tisagenlecleucel for Diffuse Large B-Cell Lymphoma. *Value Health* **25**, 1344–1351 (2022).
8. Siegler, E. L. & Kenderian, S. S. Neurotoxicity and Cytokine Release Syndrome After Chimeric Antigen Receptor T Cell Therapy: Insights Into Mechanisms and Novel Therapies. *Front. Immunol.* **11**, 2 (2020).
9. Santomasso, B. D. et al. Management of Immune-Related Adverse Events in Patients Treated With Chimeric Antigen Receptor T-Cell Therapy: ASCO Guideline. *J Clin Oncol* **39**, 3978–3992 (2021).
10. Xiao, X. et al. Mechanisms of Cytokine Release Syndrome and Neurotoxicity of CAR T-cell Therapy and Associated Prevention and Management Strategies. *J Exp Clin Cancer Res* **40**, 367 (2021).
11. Majzner, R. G. & Mackall, C. L. Tumor Antigen Escape from CAR T-cell Therapy. *Cancer Discovery* **8**, 1219–1226 (2018).
12. Larson, R. C. & Maus, M. V. Recent Advances and Discoveries in the Mechanisms and Functions of CAR T cells. *Nat Rev Cancer* **21**, 145–161 (2021).
13. Lanier, L. L. Up on the Tightrope: Natural Killer Cell Activation and Inhibition. *Nat Immunol* **9**, 495–502 (2008).
14. Zhang, L., Meng, Y., Feng, X. & Han, Z. CAR-NK Cells for Cancer Immunotherapy: From Bench to Bedside. *Biomarker Research* **10**, 12 (2022).
15. Abel, A. M., Yang, C., Thakar, M. S. & Malarkannan, S. Natural Killer Cells: Development, Maturation, and Clinical Utilization. *Front Immunol* **9**, 1869 (2018).
16. Liu Enli et al. Use of CAR-Transduced Natural Killer Cells in CD19-Positive Lymphoid Tumors. *New England Journal of Medicine* **382**, 545–553 (2020).
17. Wang, W. et al. Breakthrough of Solid Tumor Treatment: CAR-NK Immunotherapy. *Cell Death Discov.* **10**, 1–16 (2024).
18. Dogan, M. et al. Engineering Human MAIT Cells with Chimeric Antigen Receptors for Cancer Immunotherapy. *The Journal of Immunology* **209**, 1523–1531 (2022).

2 Strategies for T-cell activation

Jessalyn Low and Andy Kah Ping Tay

Introduction

The activation process of T cells in CAR-T cell manufacturing is a crucial process to induce T cell activation and prime them for genetic engineering before re-infusion back into the patient. An optimal activation process is integral to ensuring the efficacy of CAR transgene activation, such that the final cell product exhibits therapeutic efficacy with potent antitumor activity. T cell activation also holds key implications for downstream processes such as T cell expansion and functionality, which are critical attributes to the success of both the manufacturing process and therapeutic efficacy post-infusion.

In principle, T cells require at least two signals for ex vivo activation – T cell receptor (TCR) engagement (Signal 1) and costimulatory signals, most notably CD28 (Signal 2). Signal 1 triggers T cell activation; however, Signal 1 alone will not induce full activation but rather an anergy state without the presence of costimulation, i.e., Signal 2. Signal 2 is essential to promote survival, clonal expansion, and differentiation. In ex vivo conditions, T cell activation is most commonly achieved via anti-CD3 and CD28 stimulation. Cytokine signaling (Signal 3), such as IL-12 and type I IFN, is also important to support T cell expansion and effector functions. These signals mimic interactions encountered by T cells from antigen-presenting cells (APCs) during in vivo activation.

Strategies for T cell activation

Biochemical strategies

Here, we discuss the main approaches that target TCR/CD28 stimulation for T cell activation.

Activation with antibody-coated beads

The use of magnetic beads coated with anti-CD3 and anti-CD28 antibodies is the gold standard protocol that most cell therapy manufacturers are using. These beads act as artificial APCs (aAPCS), where the immobilized antibodies provide primary and costimulatory signals upon interaction with the T cells (Figure 2.1a). Additionally, the magnetic properties allow for the removal of beads from the cell products via magnetic separation at the end of the cell manufacturing process to ensure a pure cell product.

To date, there have been several Good Manufacturing Practices (GMP) clinical-grade magnetic beads available for commercial manufacturing of CAR-T cell therapies.

DOI: 10.1201/9781032660752-2

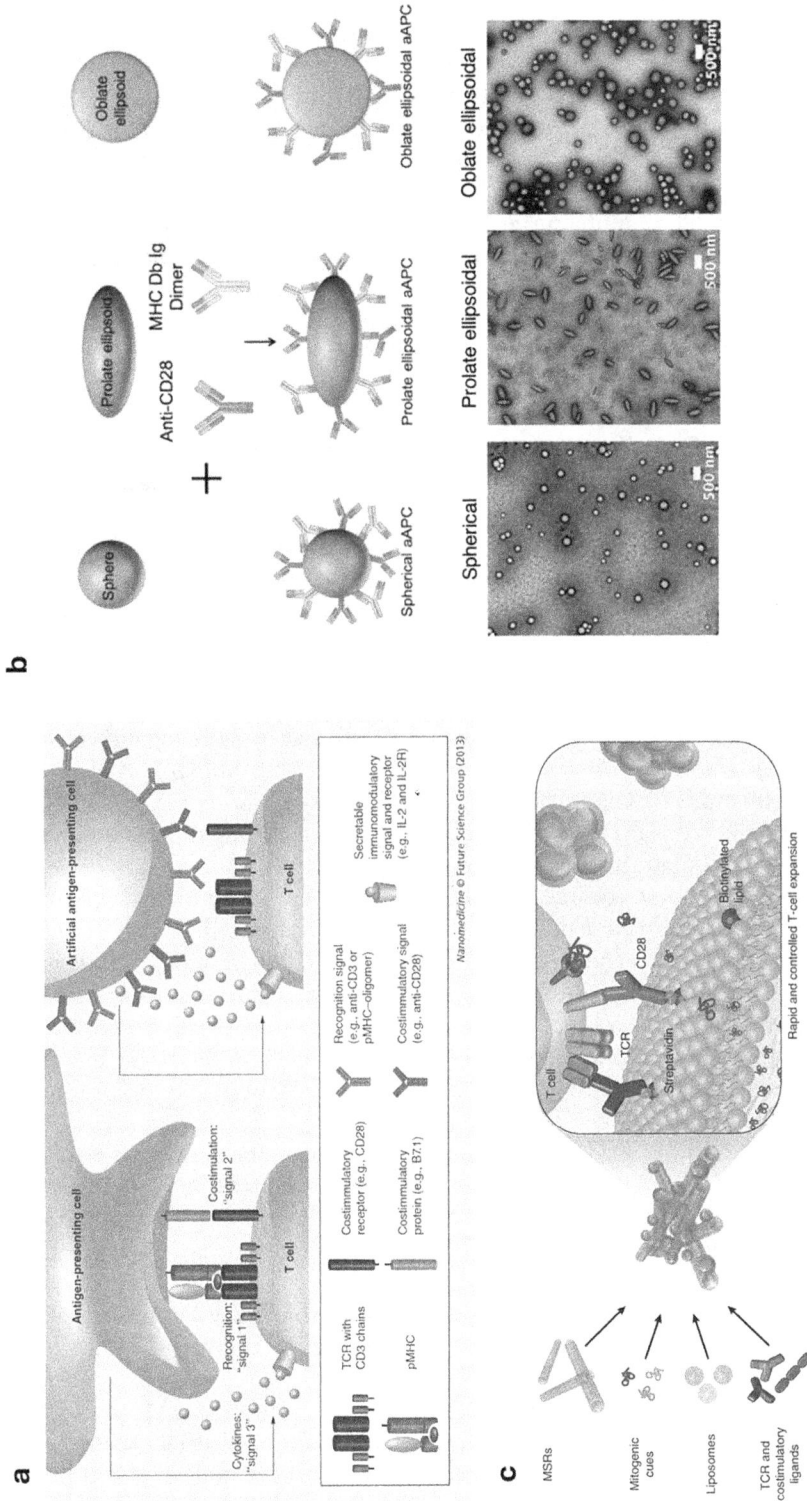

Figure 2.1 aAPCs for T cell activation (a) APC and aAPC interacting with T cells. Three signals – TCR recognition, costimulation, and cytokine signaling influence T cell activation and fate. Figure adapted from Sunshine and Green (2013)[10] with permission. (b) Schematic and TEM images of spherical and anisotropic PLGA aAPCs, fabricated by the single emulsion technique and film stretching technique, respectively. Nanoparticles were conjugated with gp100-loaded MHC Db IgG dimer and anti-CD28 monoclonal antibody to form nano-APCs. Figure adapted from Ben-Akiva et al. (2023)[4] with permission. (c) Schematic of APC-mimetic scaffolds (APC-ms). APC-ms consist of high-aspect-ratio silica microrods, which are loaded with soluble mitogenic cues and coated with liposomes of specific compositions to create supported lipid bilayers. Figure adapted from Zhang et al. (2020)[8] with permission.

For instance, CTS™ (Cell Therapy Systems) Dynabeads™ CD3/CD28, manufactured by Thermo Fisher Scientific, was used by Novartis in the manufacturing of Kymriah – the first FDA-approved CAR-T therapy. These CTS Dynabeads are 4.5-µm superparamagnetic beads covalently bound with anti-CD3 and anti-CD28 antibodies.

Utilization of magnetic beads for T cell activation has multiple advantages. One main advantage is the precision of control due to the uniform physiochemical properties of the beads, ensuring consistent and reproducible results. In addition, the use of beads is easily scalable, therefore simplifying manufacturing workflows and supporting the move toward large-scale CAR-T manufacturing processes. Owing to its versatility and scalability, the use of magnetic beads therefore remains the workhorse for T cell activation in the cell manufacturing process. The importance of using beads as an approach for T cell activation has also been recognized by biotechnology companies, which have actively endeavored to optimize the design of these beads as a means to optimize the cell manufacturing process.

In 2023, Thermo Fisher Scientific announced its next-generation CTS Dynabeads products – the CTS™ Detachable Dynabeads™ CD3/CD28 Kit. These Dynabeads retain the core technology of their CTS Dynabeads predecessor, while possessing an active release mechanism. This allows cell therapy manufacturers to enable the active release of these beads from the target cells at any point in time during the cell manufacturing process, bringing about significant impacts to the potential of optimizing the manufacturing workflow by controlling the duration of T cell activation and shortening the manufacturing process.

It should also be noted that while the beads are typically cell-sized as a way to mimic APCs in vivo, nanobeads have also been manufactured, in particular by Miltenyi Biotech – the MACS® GMP T Cell TransAct™. These are colloidal polymeric nanomatrices conjugated to humanized CD3 and CD28 agonists. Due to their nanoscale structure, the beads are soluble and can be removed via simple washing, therefore eliminating the de-beading step and streamlining the cell manufacturing process.

It is also important to take some considerations into account in the use of beads for T cell activation. For instance, bead-cell density must be carefully optimized, as they could influence the extent of CD3/CD28 engagement and consequently their activation. It is also essential to ensure that bead-based activation is paired with a thorough process to remove excess beads from the prior cell product prior to further downstream processing to avoid potential adverse safety effects associated with the presence of the beads in the final cell product. As mandated by the FDA, the final CD3/CD28 bead number in the final cell product must be below 100 per 3 million cells.

In recent research, more studies are looking at how the design of the activation beads can be optimized. Because these aAPCs are designed to mimic DCs to form an immunological synapse, particle design parameters such as particle shape, size, stiffness, and material are important as they can influence the cell-cell interface and therefore T cell activation.[1]

For instance, as natural APCs are not perfectly spherical in shape, studies have been conducted to understand the role of aAPC geometry in their interaction with T cells. It has been demonstrated that non-spherical ellipsoidal aAPCs were able to induce stronger and more efficient antigen-specific T cell responses compared to spherical aAPCs, both for microscale aAPCs[2] and nanoscale aAPCs.[3] An anisotropic shape also increases the radius of curvature and surface area for T cell contact[4] (Figure 2.1b). To mimic the softer mechanical properties of aAPCs, studies have also looked at replacing the conventional, rigid bead with a mechanically soft elastomer using polydimethylsiloxane (PDMS), where results showed a resultant extended proliferative phase.[5]

Activation with soluble monoclonal antibodies

Soluble antibodies can also be used for T cell activation. One typical approach is to use the soluble anti-CD3 monoclonal antibody OKT3 in the presence of interleukin-2 (IL-2), or anti-CD3 and anti-CD28 antibodies. For instance, GMP ImmunoCult™ Human CD3/CD28 T Cell Activator from STEMCELL Technologies consists of soluble antibody complexes that bind CD3 and CD28 cell surface ligands.

Unlike the use of activation beads that allow for localized focal signals at the point of bead-cell contact, activation with soluble antibodies does not promote TCR clustering. This results in short-lived signaling effects and slower expansion rates. However, activation via soluble antibodies could in turn then yield less-differentiated T cells with a higher proportion of naïve T cells, which may be functionally more favorable once re-infused back into the patient with increased homing and proliferative capabilities. This consideration is particularly important for CD8 cells, which may have been previously antigen-stimulated, and in cases where central lymphoid trafficking of adoptively transferred CD8 cells will be favorable.[6]

Activation with APCs and aAPCs

In vivo, APCs such as dendritic cells (DCs) endogenously activate T cells. While using APCs for T cell activation ex vivo may be most physiologically relevant and closely simulate activation in vivo, there are critical challenges that hamper the use of APCs in clinical practice. For instance, donor-to-donor variation in the ability of APCs to activate specific T cell populations may affect the consistency of the final cell product. APCs may also be of limited availability, particularly in autologous cell therapy where the donor will be from a patient.

With the unreliability of APCs limiting their use for T cell activation by cell manufacturers, aAPCs have been developed. aAPCs are genetically engineered cell lines that aim to mimic natural APCs through the expression of specific antigens and costimulatory molecules. A key example is the K562 cell line. K562 cells are derived from the myelogenous leukemia cells and lack major histocompatibility complex (MHC) class I and II molecules, making them therefore suitable to present exogenously introduced antigens to the T cells without inducing alloreactivity.

While the aAPCs allow for T cell activation in a more controllable way compared to natural APCs, like other cell-based therapies, there is still a risk of immunogenicity post-infusion into the patient that may trigger adverse reactions. K562-aAPCs must be inactivated prior to their use as feeder cells for T cell activation to avoid pathogenicity, typically by gamma irradiation. As such, K562 cells are not detected post-T cell expansion. However, there have still been reservations about using K562 in CAR-T cell manufacturing due to them being derived from a malignant cell line. In addition, compared to beads, storage is a concern as aAPCs may have lower stability and a shorter shelf-life, which may impact the cell manufacturing process and hinder the scaling up of these processes.

Besides the use of cell lines, biomaterials have been leveraged to develop scaffolds that mimic APCs closer. Compared to other techniques like beads, these scaffolds are able to present larger sets of costimulatory cues that may benefit the generation of highly functional therapeutic T cells[7,8] (Figure 2.1c). For instance, the development of APC-mimetic scaffolds, which consist of a fluid lipid bilayer supported by mesoporous silica micro-rods loaded with IL-2, has been reported. The lipid bilayer presents membrane-bound cues for TCR engagement and costimulation, while the micro-rods allow for sustained release of soluble paracrine cues, emulating how these cues are presented on an APC plasma membrane.[7] An artificial lymph node-biomimetic scaffold to mimic the porous bulk structure and bio-functional stimulatory signals of the lymph node has also been reported. Here, the scaffold was constructed using porous microspheres functionalized with anti-CD3 and anti-CD28 antibodies, as well as cytokines IL-7 and IL-15.[9]

Mechanical strategies

While many existing strategies for T cell activation focus on using TCR/CD28 stimulation, it should be noted that T cell activation is not only a biochemical process but also mechanical in nature (Figure 2.2a). As such, there has been increasing interest in elucidating the mechanosensitive nature of T cells to better engineer platforms for T cell activation.

During T cell activation, adhesion first occurs between the T cell and APC as they initially interact with each other via surface receptors like ICAM-3, following which the TCR recognizes and binds to the antigen presented by the MHC molecules on the APC. Costimulatory molecules such as CD28 on the T cell bind to CD80/CD86 expressed on the APCs, resulting in a physical contact between the T cell and APC. This physical contact in turn generates mechanical forces, leading to a slew of conformational changes in proteins and signaling molecules and ultimately forming the immunological synapse (IS), a highly organized structure between the T cell and APC that allows for communication between the two cells.

Notably, mechanical forces acting on the T cells bound to APCs via the TCR complex, but not other surface receptors, directly initiate signaling in T cells, suggesting that T cells are mechanically sensitive during TCR engagement and that the TCR behaves as a mechanosensor.[11] While the strength of the TCR signal is important, physical properties of the antigen-presenting surface such as the surface stiffness, as well as spatial structure in the micrometer and nanometer scales, have also been demonstrated to influence the mechanical forces acting on the T cell and thus influencing T cell activation.[12] Wide ligand spacing and a soft surface result in weak mechanical forces and low spreading of T cells on the activating surface, resulting in weak T cell activation. Conversely, dense ligands on hard surfaces lead to strong mechanical forces and consequently stronger T cell activation.

Activation with biomaterial scaffolds

The use of biomaterials is an attractive approach to leverage on the mechanosensitive nature of T cells for T cell activation, owing to their tunable structural and physicochemical properties. While still in its infancy, more research is being done to develop biomaterials capable of modulating the mechanical forces applied to T cells for T cell activation.

One key approach to directing T cell activation is via a combinatory approach by tethering stimulatory cues like anti-CD3 and anti-CD28 antibodies to biomaterial scaffolds whose mechanical properties are highly tunable.

In particular, substrate stiffness is a mechanical property that is often exploited, due to their established involvement in regulating T cell activation and proliferation[13] (Figure 2.2b,c). For instance, antibody-coated 3D scaffold matrices with tunable stiffness and

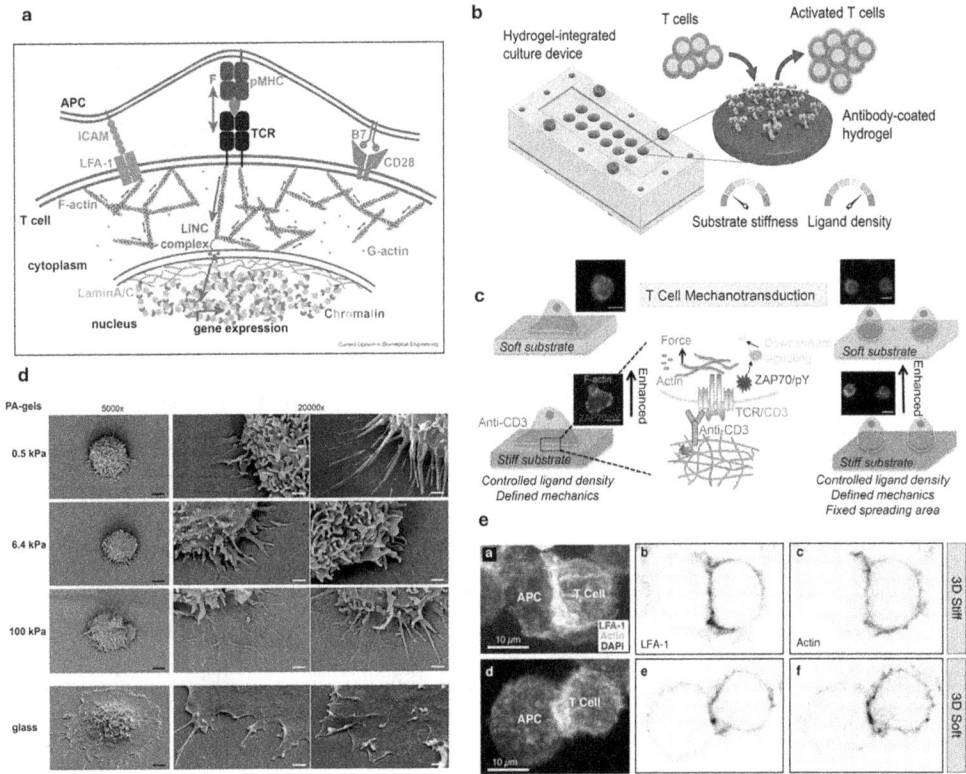

Figure 2.2 Mechanical activation of T cells. (a) Mechanotransduction from activating surface to nucleus. T cells are able to sense mechanical forces applied through their TCR. Figure adapted from Aramesh et al. (2019)[12] with permission. (b) A hydrogel-integrated culture device designed to study the physicochemical stimulation of T cells. Substrate stiffness and ligand density were tuned by concentrations of the hydrogel cross-linker and antibody in the coating solution, respectively. Figure adapted from Chin et al. (2020)[20] with permission. (c) Micropatterned soft hydrogel of different stiffness, with arrays of anti-CD3 microdots designed to study hydrogel-T cell contact and downstream events of TCR activation. Figure adapted from Zhang et al. (2021)[17] with permission. (d) SEM image of T cells interacting with poly-acrylamide gel substrates of different stiffness. T cells form small protrusions that spread more on stiffer substrates. Figure adapted from Saitakis et al. (2017)[16] with permission. (e) 3D confocal fluorescence microscopy of immune synapse size formed by T cells and APCs on 3d scaffolds of different stiffness. Immune synapses of T cells formed in stiff matrices were larger than those in soft matrices. Figure adapted from Majedi et al. (2020)[14] with permission.

consistent microporosity have been developed, where an augmentation of T cell activation in stiffer 3D matrices has been observed.[14] In another study, hyaluronic acid (HA)-based hydrogels with conjugated anti-CD3 and anti-CD28 antibodies were developed with different stimulatory ligand density and stiffness, which were then shown to influence T cell activation and control the phenotype of the cell product.[15]

Other studies have also reflected the difference in morphology of T cells and immune synapse size on substrates of different stiffness, further demonstrating the involvement of

TCR as a mechanosensor during the activation process[14,16] (Figure 2.2d, e). Force exertion on TCR induces phosphorylation of CD3 signaling chains, as associated with conformational changes in the CD3 complex. The phosphorylated tyrosine-based activation motifs can then bind to kinase ZAP-70 and activate downstream signaling events. In a study it was shown that ZAP-70 phosphorylation level appeared to be dependent on the hydrogel stiffness.[17]

Viscoelasticity, which describes the material's time-dependent mechanical response to applied strains and is distinct from stiffness, has also been demonstrated to influence T cell activation. By altering the viscoelasticity of the surrounding collagen type 1-based ECM model, it was shown that ECM viscoelasticity regulates T cell phenotype and function via the activator-protein-1 (AP-1) signaling pathway, which is a critical regulator of T cell activation and fate. By tuning ECM viscoelasticity, functionally different T cell populations were generated – more viscous hydrogels with weakly crosslinked collagen fibers resulted in memory-like T cells, while more elastic hydrogels with strongly cross-linked collagen fibers resulted in effector-like T cells.[18]

Activation with mechanical forces and stresses

Exposing T cells to external mechanical stresses can also be used to aid the T cell activation process.

For example, it has been shown that by applying a uniform fluid shear stress to T cells using a cone-and-plate viscometer system, in combination with soluble and bead-bound CD3/CD28 antibodies, activation of signaling proteins essential for T cell activation like zeta-chain-associated protein kinase-70 (ZAP-70) and AP-1 is increased. This is attributed to the activation of the mechanosensitive ion channel Piezo1, which promotes calcium influx and consequently drives the expression of T cell activation markers.[19]

Future perspectives

T cell activation is a critical step in CAR-T cell manufacturing, and various strategies employed have significantly advanced the field of CAR-T cell therapy. Current approaches like the use of beads, APCs, and soluble antibodies have been employed in a wide range of clinical trials and existing FDA-approved CAR-T cell therapies, demonstrating their effectiveness in generating cell products for CAR-T cell therapy. With recent successes in CAR-T cell therapy, there has also been a greater interest in pushing for wider accessibility of CAR-T therapy by enhancing the scalability and cost-effectiveness of manufacturing processes. For instance, more biotechnology companies are actively optimizing and innovating activation methods to streamline the cell manufacturing process. Deeper understanding into the mechanisms of T cell activation has also brought about much active research to develop novel approaches such as through the use of biomaterials as well as mechanical activation.

According to GlobalData's database, in 2023, there were over 210 companies, including pharmaceutical companies and startups, engaged in the development and application of in vitro T cell activation. For instance, Akadeum Life Sciences is an emerging startup that has developed a Human T Cell Selection, Activation, and Expansion Kit. This is based on its novel microbubble technology, which helps to combine T cell selection, activation, and expansion steps into a single workflow. By exploiting the natural buoyancy of microbubbles that get bound to the antibody-labeled cells, T cell activation and expansion take place at the top of the culture vessel. Following the proliferation of activated T cells, daughter cells dissociate from the microbubbles and settle at the bottom, thus being separated from the activation signals and preventing overstimulation.

Nanotein Technologies, another startup, also works on developing novel cell activation and expansion reagents. Nanotein's NanoSpark™ STEM-T Soluble T Cell Activator is soluble in nature and contains self-assembling protein nanoparticles with functional antibodies bound to the surface through high-affinity non-covalent interactions, which aim to preserve stemlike CD8+ T cells during T cell expansion. Nanotein's NanoSparkTM EVEN-T Soluble T Cell Activator reagent kit further allows the user to modulate the CD4+:CD8+ ratio in a dose-dependent manner during T cell expansion.

It should also be noted that while this chapter focuses on strategies for T cell activation, studies have also looked at the potential of using non-activated CAR-T cells to eliminate the process of ex vivo expansion associated with loss of potency, while also reducing the duration of the manufacturing process to achieve shortened vein-to-vein time, which could be potentially useful for patients with rapidly progressive disease. In a recent study, functional CAR-T cells were generated in just 24 hours, unlike the conventional cell manufacturing process that typically takes 9 to 14 days. This was achieved by exploiting the unique ability of lentiviral vectors to transduce non-activated quiescent T cells, therefore eliminating the activation step in the manufacturing process and minimizing ex vivo manipulation. One consideration, however, is that the T cell subsets found to be optimal for activated T cells may not necessarily be the same as when using non-activated T cells in adoptive immunotherapy; as such, CAR design for quiescent T cells should be optimized. Combining T cell subsets with their preferred costimulatory domain may be more beneficial for immediate effector function and long-lasting engraftment.[21]

Regardless of the approach taken to activate T cells, proper optimizations to the activation process must be taken to ensure optimal activation duration and strength. Low levels of activation signals could lead to suboptimal activation, while high levels of activation signals could conversely lead to activation-induced cell death (AICD). The process of activation can furthermore be tailored to favor the preferential activation and expansion of certain cell subsets due to their differing activation thresholds. For instance, under a specific activation signal, while it may cause AICD of memory T cells, it could be the optimal activation signal for the activation and proliferation of naïve T cells, therefore favoring the presence of naïve T cells in the final cell product. Under a lower activation signal, memory T cells may be activated, but insufficient for naïve T cells, therefore favoring the presence of memory T cells. Such differences may be exploited to then achieve the desired phenotype of the CAR-T cell population, which may have significant implications for functional attributes such as persistence and antitumor activity, potentially enhancing therapeutic efficacy and enabling patient-specific personalized medicine.

References

1. Wang, C., Sun, W., Ye, Y., Bomba, H. N. & Gu, Z. Bioengineering of Artificial Antigen Presenting Cells and Lymphoid Organs. *Theranostics* 7, 3504–3516 (2017).
2. Sunshine, J. C., Perica, K., Schneck, J. P. & Green, J. J. Particle Shape Dependence of CD8+ T Cell Activation by Artificial Antigen Presenting Cells. *Biomaterials* 35, 269–277 (2014).
3. Meyer, R. A. et al. Biodegradable Nanoellipsoidal Artificial Antigen Presenting Cells for Antigen Specific T-Cell Activation. *Small* 11, 1519–1525 (2015).
4. Ben-Akiva, E. et al. Shape Matters: Biodegradable Anisotropic Nanoparticle Artificial Antigen Presenting Cells for Cancer Immunotherapy. *Acta Biomaterialia* 160, 187–197 (2023).

5. Lambert, L. H. et al. Improving T Cell Expansion with a Soft Touch. *Nano Lett.* **17**, 821–826 (2017).
6. Li, Y. & Kurlander, R. J. Comparison of Anti-CD3 and Anti-CD28-coated Beads with Soluble Anti-CD3 for Expanding Human T Cells: Differing Impact on CD8 T Cell Phenotype and Responsiveness to Restimulation. *J Transl Med* **8**, 104 (2010).
7. Cheung, A. S., Zhang, D. K. Y., Koshy, S. T. & Mooney, D. J. Scaffolds That Mimic Antigen-presenting Cells Enable ex Vivo Expansion of Primary T Cells. *Nat Biotechnol* **36**, 160–169 (2018).
8. Zhang, D. K. Y., Cheung, A. S. & Mooney, D. J. Activation and Expansion of Human T Cells Using Artificial Antigen-presenting Cell Scaffolds. *Nat Protoc* **15**, 773–798 (2020).
9. Liao, Z. et al. Lymph Node-biomimetic Scaffold Boosts CAR-T Therapy Against Solid Tumor. *National Science Review* **11**, nwae018 (2024).
10. Sunshine, J. C. & Green, J. J. Nanoengineering Approaches to The Design of Artificial Antigen-Presenting Cells. *Nanomedicine* **8**, 1173–1189 (2013).
11. Li, Y.-C. et al. Cutting Edge: Mechanical Forces Acting on T Cells Immobilized via the TCR Complex Can Trigger TCR Signaling. *The Journal of Immunology* **184**, 5959–5963 (2010).
12. Aramesh, M., Stoycheva, D., Raaz, L. & Klotzsch, E. Engineering T-cell Activation for Immunotherapy by Mechanical Forces. *Current Opinion in Biomedical Engineering* **10**, 134–141 (2019).
13. O'Connor, R. S. et al. Substrate Rigidity Regulates Human T Cell Activation and Proliferation. *The Journal of Immunology* **189**, 1330–1339 (2012).
14. Majedi, F. S. et al. T-cell Activation is Modulated by the 3D Mechanical Microenvironment. *Biomaterials* **252**, 120058 (2020).
15. Hickey, J. W. et al. Engineering an Artificial T-Cell Stimulating Matrix for Immunotherapy. *Advanced Materials* **31**, 1807359 (2019).
16. Saitakis, M. et al. Different TCR-Induced T Lymphocyte Responses are Potentiated by Stiffness with Variable Sensitivity. *eLife* **6**, e23190 (2017).
17. Zhang, J. et al. Micropatterned Soft Hydrogels to Study the Interplay of Receptors and Forces in T Cell Activation. *Acta Biomaterialia* **119**, 234–246 (2021).
18. Adu-Berchie, K. et al. Generation of Functionally Distinct T-Cell Populations by Altering the Viscoelasticity of Their Extracellular Matrix. *Nat. Biomed. Eng* **7**, 1374–1391 (2023).
19. Hope, J. M. et al. Fluid Shear Stress Enhances T Cell Activation through Piezo1. *BMC Biology* **20**, 61 (2022).
20. Chin, M. H. W., Norman, M. D. A., Gentleman, E., Coppens, M.-O. & Day, R. M. A Hydrogel-Integrated Culture Device to Interrogate T Cell Activation with Physicochemical Cues. *ACS Appl. Mater. Interfaces* **12**, 47355–47367 (2020).
21. Ghassemi, S. et al. Rapid Manufacturing of Non-activated Potent CAR T Cells. *Nat Biomed Eng* **6**, 118–128 (2022).

3 Genetic tools to engineer CAR-T cells

Lessons from clinical studies

Yu Yang Ng and Andy Kah Ping Tay

Introduction

Cancer immunotherapy harnesses the body's own immune system to combat cancer. Recognized as a scientific breakthrough in 2013, this therapy has shown promising antitumour efficacy in recent years. Cancer immunotherapies fall into categories such as immune checkpoint inhibitors (ICIs), adoptive cell therapies (ACTs), and tumour vaccines.[1,2]

ACT stands out as a trailblazing approach in immunotherapy, notable for its broad applicability and the swift advancements and therapeutic milestones it has achieved. ACT has been developed through various iterations, beginning with autologous tumour-infiltrating lymphocyte (TIL) therapy,[3,4] and advancing to antigen-specific endogenous T cell therapy, leading to the creation of chimeric antigen receptor (CAR) and T cell receptor (TCR)-T cell therapies.[5-7] CAR-T cells have the unique ability to identify antigens on the extracellular membrane[8] directly, bypassing the constraints associated with the major histocompatibility complex (MHC), which is one of the main considerations for TIL and TCR-T cell therapies.[9] Till date, seven CAR-T cell therapies have been approved by the Food and Drug Administration in the United States of America,[10,11] and one CAR-T cell therapy has been approved for use in China.[12] Such progress has led to exceptional progress where patients with B cell malignancies have complete response (CR) rates of 60%–70% after CAR-T cell treatment.[13-16]

Concurrently, development of the manufacturing process for ACT has evolved from the expansion of T cell populations derived from tumour specimens,[3,4,17] into the use of genetic engineering to modify T cells to obtain engineered T cell populations recognizing specific tumour targets through either the use of CAR[11,18] or TCR.[9] As such, choosing an appropriate methodology to genetically engineer T cells plays a pivotal role in determining the clinical efficacy of CAR-T cell products. Key factors to consider in choosing a suitable engineering tool include transgene expression stability and efficiency, genetic cargo capacity, safety, and the potential for rapid, cost-effective production scale-up for clinical translation. In this chapter, we embark on an exploration of gene editing techniques, such as genome-editing tools, viral vectors, and transposon systems, as well as RNA-based genetic engineering of CAR-T cells, which has experienced a resurgence, particularly with the emergence of mRNA-based vaccines during the COVID-19 pandemic. We will also discuss how each genetic tool is actively being adopted to modify CAR-T cells in ongoing clinical studies. Finally, we will provide insights into the future direction of genetic engineering for T cell immunotherapy.

DOI: 10.1201/9781032660752-3

Viral-based genetic modification

Overview of viral vectors for CAR-T cell modification

The use of viral vectors is a cornerstone in the genetic engineering of CAR-T cells, playing a pivotal role in modifying and enhancing their therapeutic potential. Gamma-retroviral and lentiviral vectors are commonly employed in CAR-T cell manufacturing attributed to their effectiveness in facilitating direct delivery and sustained expression of a transgene.[19] Additionally, they possess a packaging capacity of up to 9 kb. One advantage offered by lentiviral vectors is their ability to infect both dividing and non-dividing cells,[20] thereby enhancing their versatility in modifying CAR-T cells. So far, all seven CAR-T cell therapies that have been approved for use are made using these types of viral vectors.[10,12]

Genomic features of the viral vectors

Viral vectors used to genetically engineer CAR-T cells are primarily derived from the Retroviridae family. These viruses are enveloped, single-stranded RNA viruses and can be categorized as simple or complex based on their genome organization. Gamma-retroviruses are examples of simple retroviruses, while lentiviruses, such as HIV-1, are complex retroviruses[20] (Figure 3.1a).

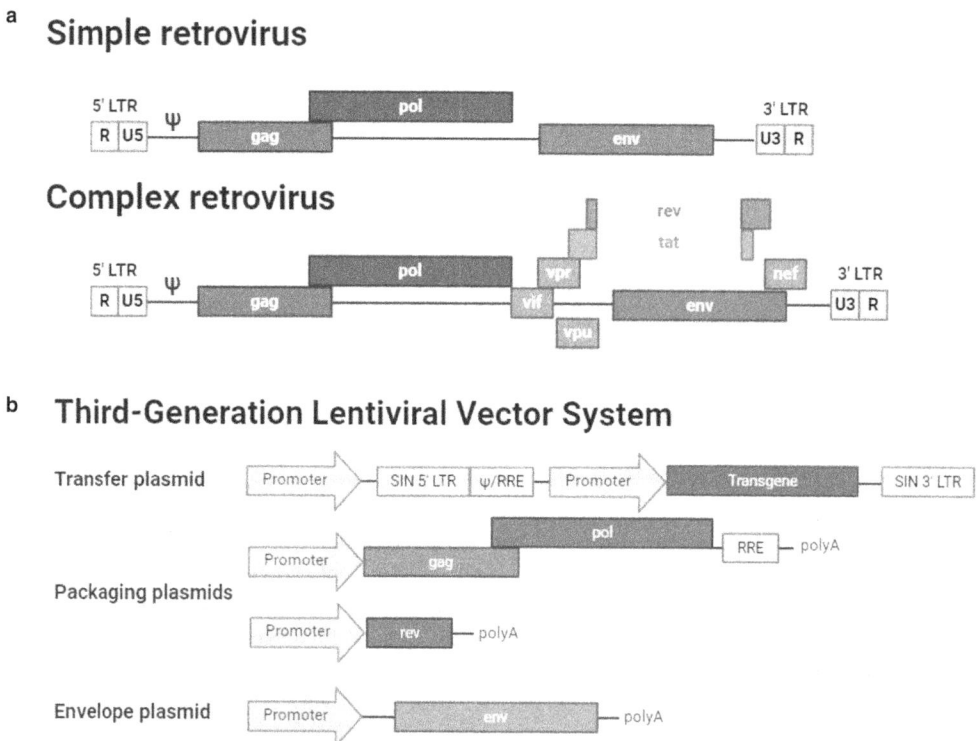

a   Simple retrovirus

Complex retrovirus

b   Third-Generation Lentiviral Vector System

Figure 3.1 Genomic features of viral vectors. (a) Genome of a simple retrovirus (top) and a complex retrovirus (bottom) showing coding regions and regulatory elements. (b) Structure of recombinant lentiviral vectors and packaging plasmid of a third-generation lentiviral vector system. Created with BioRender.com.

Structurally, the retroviral genome is flanked by 5'- and 3' long terminal repeats (LTRs), which are essential for viral genome replication. The LTR comprises a common shared R sequence and virus unique U5, U3 sequences, located at the 5' and 3' ends, respectively. Each of these sequences serves an important function: R is required for reverse transcription and replication, U5 supports the initiation of reverse transcription, and U3 contains enhancer and promoter elements. The packaging signal known as psi (Ψ) is important for the encapsidation of the viral genome, ensuring high-efficiency packaging of the transcripts into virions. Finally, retroviruses share essential core protein genes like *gag*, *pol*, and *env*. The *gag* gene is responsible for synthesizing the capsid proteins, which are the shell that protects the viral genetic material. The *pol* gene provides the enzymes needed for the virus to copy its genetic material and integrate it into the host's genome, a crucial step for the virus to replicate within host cells. Finally, the *env* gene produces the viral envelope glycoprotein, which helps the virus enter and infect host cells by binding to their surface.[20]

In most viral vectors, the endogenous *env* gene is usually replaced by a foreign *env* gene encoding vesicular stomatitis virus glycoprotein (VSV-G).[21,22] This process is known as pseudotyping, where production of viral particles is combined with a foreign viral envelope protein. The VSV-G binds to the low-density lipoprotein (LDL) receptor,[23] which is broadly expressed on various cell types, including activated T cells,[24] thus facilitating successful transduction.

Safety designs in the retroviral and lentiviral systems

In the current era of clinical trials, insertional mutagenesis is recognized as the primary limitation of retrovirus-based vectors, including lentiviral vectors (LV).[25] It is also crucial to remember that the development of replication-competent viruses raises significantly greater safety concerns, as the consequences can be severe and may extend beyond the individual receiving the treatment. As a result, US FDA demands 15 years of follow-up for every patient who has either participated in clinical trials involving CAR-T cell or has received approved CAR-T products.[26]

To prevent the formation of replication-competent species, the conversion of lentiviruses into viral vectors involves a split genome design by segregating sequences vital for viral packaging and production while removing those encoding unnecessary viral proteins. Through several modifications, the third-generation lentiviral vectors are comprised of four distinct components[19] (Figure 3.1b): (i) the *rev* packaging plasmid, producing proteins responsible for exporting the viral genome, (ii) *gag-pol* packaging plasmid, producing structural proteins and enzymes; (iii) an *env* plasmid encoding VSV-G, aiding virus entry into cells; and (iv) the transfer plasmid containing the transgene with a self-inactivating (SIN) long terminal repeat (LTR) configuration. This configuration includes a deletion in the U3 region of the 3' LTR, which contains promoter/enhancer elements. This deletion is then replicated at the 5' LTR during reverse transcription, reducing the risk of nearby gene activation. This modification enhances the safety profile of the lentiviral vector by minimizing potential off-target effects.[19,20]

Outlook of viral-based engineering of CAR-T cells

As of now, there are a total of 364 registered ongoing clinical trials involving the use of viral-transduced CAR-T cells, with 40 of them registered in 2023 (information retrieved from www.clinicaltrials.gov). This number suggests that the utilization of the lentiviral platform for CAR-T cell engineering remains highly sought after, attributed to the recognized clinical successes and ongoing commitments to enhance the manufacturing of viral vectors. A current focal point in the field is the repurposing of viral vectors for in vivo CAR-T cell transfection and generation,[27] which could simplify the intricate manufacturing process of ex vivo CAR-T cell production. Several preclinical studies have demonstrated the feasibility of in vivo CAR-T cell transfection, and biotech companies like Umoja Biopharma,[28] Exuma Biotech,[29] and Interius BioTherapeutics[30] have active pipelines for their in vivo CAR-T cell transfection technology utilizing viral vectors. With a predicted outgrowth in the lentiviral vector market at a compounded annual growth rate (CAGR) of 18.5% from 2021 to 2030, it is likely that the use of viral vectors in the engineering of T cells will remain an attractive option in the future.

Viral-free genetic engineering

Rationale for non-viral engineering approach

CAR-T cells generated using viral vectors have shown remarkable effectiveness in antitumour efficacy in both clinics and research settings; however, a major downside with the use of these vectors lies in the high costs required to manufacture clinical-grade certified products.[31] Additionally, in November 2023, the FDA announced its investigation into reported cases of CAR-positive T cell lymphoma in patients who had previously received CAR-T cell infusion, highlighting the inherent risk of insertional mutagenesis associated with the use of viral vectors.[32] Because of these issues, other methods to genetically modify T cells, such as delivery of mRNA, genome editing, or using the transposon system, should be considered.

Transposon system

Overview

Eukaryotic genomes comprise a plethora of repeated DNA, and certain sequences among them can relocate from one location to another with the genome. These sequences are known as transposable elements (TEs) and have been identified in all organisms, including both prokaryotes and eukaryotes.[33]

TEs fall into two main categories: retrotransposons and DNA transposons. Retrotransposons use an RNA intermediate in a copy-and-paste mechanism, while DNA transposons employ a DNA intermediate in a cut-and-paste mechanism.[33] DNA transposons typically consist of a transposase gene that is flanked by terminal inverted repeats (TIRs). The transposase binds to sequences within the ITRs, excising the transposable element from its original position and integrating it into a different location in the genome. Modern transposon-based vector systems used for genetic engineering involve separating the transposase and TIRs into two components, with the transgene cassette positioned between two TIRs in a separate vector. In the following section, we will briefly describe the Sleeping Beauty (SB) and the piggyBac (PB) transposon systems, which are mainly utilized for CAR-T cell engineering.[34]

Characteristics of the SB and PB systems

Sleeping Beauty (SB) is among the most widely used DNA class II transposons as a genetic tool. A typical SB transposon system comprises two components: the SB transposon

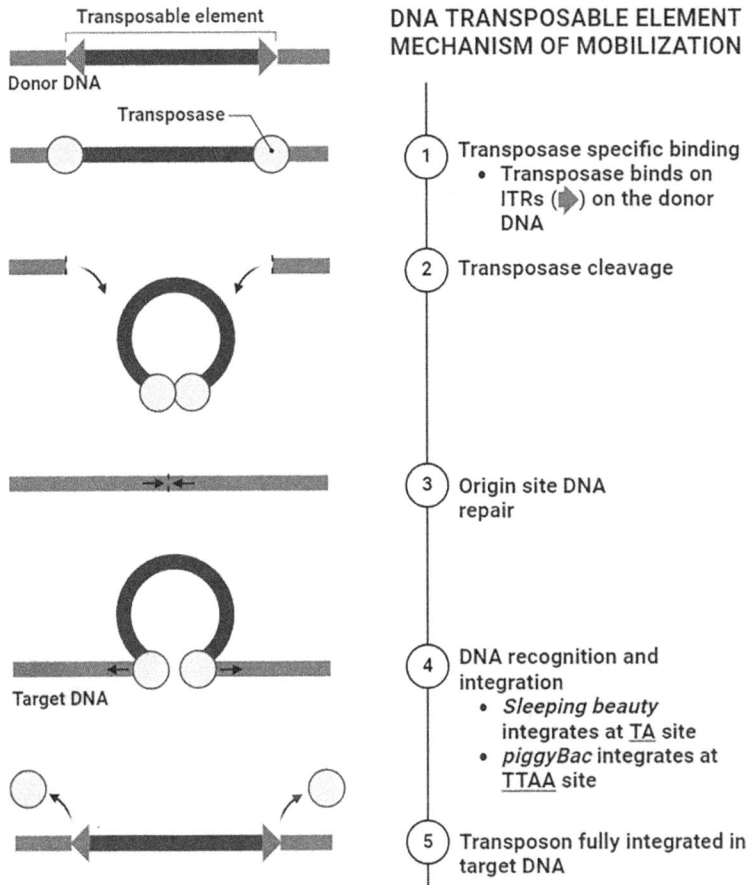

DNA TRANSPOSABLE ELEMENT MECHANISM OF MOBILIZATION

1. Transposase specific binding
   - Transposase binds on ITRs (▶) on the donor DNA

2. Transposase cleavage

3. Origin site DNA repair

4. DNA recognition and integration
   - *Sleeping beauty* integrates at <u>TA</u> site
   - *piggyBac* integrates at <u>TTAA</u> site

5. Transposon fully integrated in target DNA

Figure 3.2 Mechanism of DNA transposition using the SB or PB system. Created with BioRender.com.

vector, containing the transgene flanked by SB terminal inverted repeats (TIRs), and the SB transposase expression vector, which recognizes the TIR sequences. The mechanism of SB transposition, as shown in Figure 3.2, occurs when the SB transposase identifies the TIR sequences bordering the SB donor transposon and binds to them, causing a double-stranded break and excision of the donor transposable element. This excised element forms a complex with the transposase, enabling it to search for a suitable target site in the host genome for integration, and moves the transgene from the donor DNA to a new location within the host genome.[35]

PiggyBac (PB) is a DNA transposon originally identified in the genome of the Cabbage Looper moth (*Trichoplusia ni*). Like the SB system, the PB system consists of the PB transposase and a separate transfer plasmid carrying the donor transgene. The integration events that occur in the PB system are like those in the SB system except that the PB system prefers to integrate at genomic regions containing TTAA (nucleotide thymine thymine adenine adenine in sequential order) sites, contrasting with the SB system where TA dinucleotides are preferred.[35]

Compared to viral vectors, the transposon system offers a larger cargo capacity of approximately >100 Kb, which is useful for future applications requiring the delivery of

Table 3.1 Comparison of key features of the viral vectors and transposon system as integrating vectors

	Viral vectors		Transposon system	
	Gamma-retroviral	Lentiviral	Sleeping Beauty	piggyBac
Cargo capacity (kb)	~8	~8	>100	>100
Stability of gene expression	High	High	High	High
Cell types	Dividing cells only	Dividing and non-dividing cells	Dividing and non-dividing cells	Dividing and non-dividing cells
Integration sites	Promoter, enhancer regions of actively transcribed genes	Regions of actively transcribed genes	Close to random integration	Biased towards transcriptional start sites
Manufacturing cost	High	High	Moderate	Moderate

multiple transgenes into the genome.[33,35] Additionally, as integrating vectors, PB and SB exhibit different preferences for integration sites. In an analysis conducted by Gogol-Döring et al., the SB transposon system demonstrated the highest probability of targeting safe harbours due to its random integration profile. In contrast, the PB transposon system displayed a bias for transcriptional start sites.[36] Therefore, to minimize the risk of insertional mutagenesis, it would be more advantageous to utilize the SB transposon as a genetic tool for engineering T cells. A summary comparing viral vectors to the transposon system is shown in Table 3.1.

Status of transposon-based CAR-T cells in clinical trials

In 2011, SB-engineered CAR-T cells made a significant leap in clinical trials, signifying the inaugural use of a non-viral vector to engineer CAR-T cells.[37] The clinical trials (NCT00968760 and NCT01497184) infused CD19 CAR-T cells into patients with non-Hodgkin's lymphoma (NHL) and acute lymphoblastic leukaemia (ALL) as an adjuvant therapy following hematopoietic stem cell transplantation (HSCT). The trials yielded promising results, where the overall survival (OS) and 30-month progression-free survival (PFS) were 100% and 83%, respectively, for patients receiving autologous CAR-T infusion (NCT00968760). In the allogeneic setting (NCT01497184), 12-month PFS and OS were 53% and 64%, respectively. Patients recruited under the trial did not show signs of CAR-T cell-related toxicity such as cytokine release syndrome (CRS) or immune effector cell-associated neurotoxicity syndrome (ICANS) during the treatment.[37] As such, these studies demonstrated that the non-viral SB platform is a safe and effective tool for manufacturing therapeutic CAR-T cells. As of now, 14 clinical trials have adopted SB as a non-viral vector for CAR-T cell production in their protocols.[34,38]

A phase-I study, known as the CARTELL Study (NHMRC identifier: 1102172), conducted in Australia in 2017 introduced the PB transposon system into clinical settings. CD19 CAR-T cells were manufactured via the PB system and administered to patients with CD19+ malignancies.[39] Results released from the clinical trial indicated about 50% of the patients achieving complete remission (CR). Unexpectedly, two patients who had previously achieved CR during the treatment developed CAR-positive T cell lymphoma, leading to death of one patient. Analysis of malignant CAR-T cells from both patients revealed that there were no signs of integration of transgene into oncogenes and therefore we were not able to determine the mechanism behind the malignant transformation of the CAR-T cells.[40] While this may suggest that the PB system may not be safe, in another clinical trial (NCT03288493), where a two-year follow-up study was performed on multiple myeloma (MM) patients treated with B cell maturation antigen (BCMA) CAR-T cells manufactured using the PB system, there were no reports of malignant transformation of CAR-T cells into lymphoma.[41,42] Nevertheless, this finding reinforces the need for strict follow-up for patients receiving CAR-T cells manufactured using genetic engineering systems that involve random integration, such as the LV, SB, or the PB systems.

Genome editing

Overview of genome editing in CAR-T cells

Researchers have developed powerful tools that can precisely cut DNA at specific positions, transforming how genes can be edited across different cell types and organisms. These tools include zinc-finger nucleases (ZFNs), transcription activator-like effector nucleases (TALENs), and the clustered regulatory interspaced short palindromic repeat/ associated nuclease protein 9 (CRISPR/Cas9) system.[43] Each of these tools has allowed scientists to make targeted modifications to DNA, which is crucial for research and therapy development. Specifically, within CAR-T cell development, these genome editing tools have facilitated techniques for both gene knockout (KO) and knock-in (KI). Applications include creating universal allogeneic CAR-T cells readily available "off-the-shelf", attenuating inhibitory receptors, or inserting the CAR gene or other transgenes directly into chosen genomic sites.[34,38]

As shown in Figure 3.3, TALENs, ZFNs, and the CRISPR/Cas9 system are pivotal in enabling precise genetic engineering by introducing specific double-strand breaks (DSBs) in DNA. DSBs activate two primary pathways for repair: non-homologous end joining (NHEJ) and homology-directed repair (HDR).[43] NHEJ, the more common but error-prone method, directly reconnects the broken DNA strands. This often results in insertions or deletions that can disrupt a gene's coding sequence, effectively knocking out the gene if it induces a frameshift mutation. Conversely, HDR is a more precise, though less frequent, mechanism of repair that occurs mainly during the late S- and G2-phases of the cell cycle, after DNA replication. This phase provides an optimal setting for HDR because the sister chromatids can act as templates for repair. Using donor DNA with homology arms, HDR allows for the specific insertion of a transgene at the site of the break, achieving a targeted gene knock-in.[43]

ZFNs are pioneering tools in targeted genome editing and are created by fusing zinc fingers (ZFs) with a restriction endonuclease, usually FokI. In ZFNs, the specificity arises from the arrangement of ZF units, each of which binds to a specific triplet of DNA bases. By assembling three to six of these ZF units, ZFNs can target unique sequences within the genome, increasing precision. Because the FokI endonuclease functions as a dimer, it necessitates the assembly of two ZFNs on opposite strands of DNA for effective double-stranded cleavage. This requirement for dual binding enhances target accuracy and limits off-target effects, as both ZFNs must be designed to simultaneously engage with their specific DNA sequences in proximity. However, designing ZF arrays that can achieve high specificity is complex as the interaction between neighbouring ZFs in a sequence can

Zinc finger nucleases (ZFNs)

N-⚬⚬⚬ Fok1 -C

ZF domain

Transcription activator-like effector nucleases (TALENS)

N-⚬⚬⚬⚬⚬ Fok1 -C

C- Fok1 ⚬⚬⚬⚬⚬-N

TALE domain

CRISPR-Cas9

Cas9

PAM

Target DNA

sgRNA

Double-stranded breaks (DSBs)

Homology-directed repair (HDR) pathway

DSB

Donor DNA

Gene insertion/replacement

Non-homologous end joining (NHEJ) pathway

DSB

Gene insertion

Gene deletion

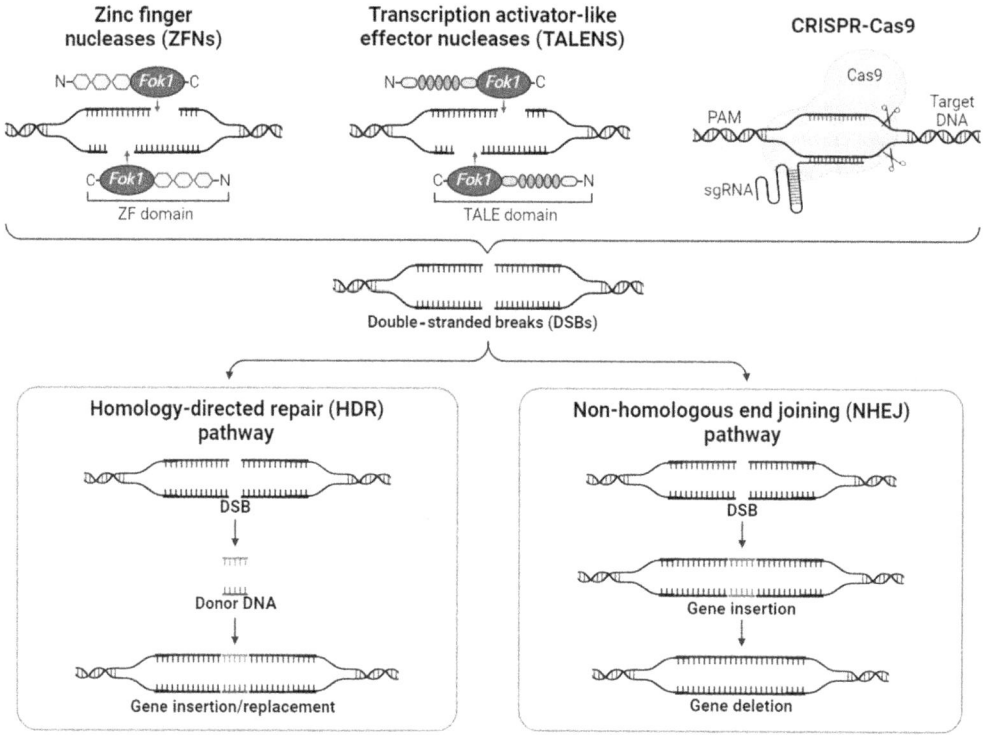

Figure 3.3 Genome editing mechanisms of the TALENs, ZFNs, and the CRISPR/Cas9 system. Created with BioRender.com.

affect the binding affinity and specificity onto the target DNA sequence, making the prediction of the overall binding characteristics of the ZFN challenging.[43]

TALENs operate on a principle of protein–DNA interaction for targeting, similar to ZFNs, but with a simpler and more straightforward design. Each transcription activator-like effector (TALE) domain within TALENs binds to a single DNA base, allowing for the assembly of an array of TALE domains to target specific DNA sequences. Unlike ZFNs, the binding of each TALE domain has no impact on neighbouring TALE binding onto the target DNA, simplifying the design and engineering process. TALE domains are fused to the FokI endonuclease, which requires dimerization to initiate DNA cleavage. For effective cleavage, pairs of TALENs must bind to complementary DNA strands near the target site, necessitating the precise placement of two TALENs. This system's design flexibility and independence of domain specificity make TALENs a more user-friendly tool for genome editing.[43]

The CRISPR/Cas9 system comprises the Cas9 nuclease and two RNA components: the trans-activating crRNA (tracrRNA) and a single-guide RNA (sgRNA). For Cas9 to function, the target DNA sequence must be followed by a specific DNA motif known as a protospacer adjacent motif (PAM), with each Cas9 variant recognizing a particular PAM sequence, such as the standard Cas9's 5'-NGG-3'. When activated, Cas9 performs DNA cleavage, potentially inducing a double-strand break with its wild-type enzyme. The CRISPR/Cas9 system operates through RNA–DNA interactions, with the RNA components guiding Cas9 to the target DNA sequence for precise cleavage.[43]

CRISPR-Cas9 offers distinct advantages over ZFN or TALEN technologies, primarily due to its flexibility and capacity for multiple gene editing. Unlike ZFNs or TALENs, which require the engineering of a unique protein for each target sequence in the genome, CRISPR-Cas9 simplifies the process significantly. With CRISPR-Cas9, the researcher only needs to vary the design of the single-guide RNAs (sgRNAs) to target different genomic regions. This streamlined approach reduces costs, labour, and the need for advanced molecular biology expertise associated with ZFN or TALEN technologies.[43]

Moreover, CRISPR-Cas9's ability to edit multiple loci simultaneously enhances its efficiency and scalability. In contrast, ZFNs and TALENs may require separate protein engineering for each target, making them less practical for large-scale or multiplex genome editing projects. Overall, CRISPR-Cas9's versatility and simplicity make it a preferred choice for a wide range of genetic engineering applications.[43]

Genome editing to deliver allogeneic CAR-T cells

Developing new CAR-T cell therapies for patients with cancer faces hurdles, especially when using their own T cells, which might not be plentiful or fully functional, especially in older or previously treated individuals. Genome editing presents a promising solution by creating "off-the-shelf" T cells from donors,[44] potentially easing manufacturing challenges and reducing costs compared to personalized therapies. However, using cells from another person raises concerns about graft-versus-host disease (GvHD), where the donor cells attack the recipient's tissues.

To address this concern, knockout of *TCR* genes, as shown in Figure 3.4, can be employed by targeting components like TCRα (*TRAC*), TCRβ (*TRBC*) constant chains, and/or beta-2 microglobulin (*B2M*), thereby reducing the likelihood of GvHD in the

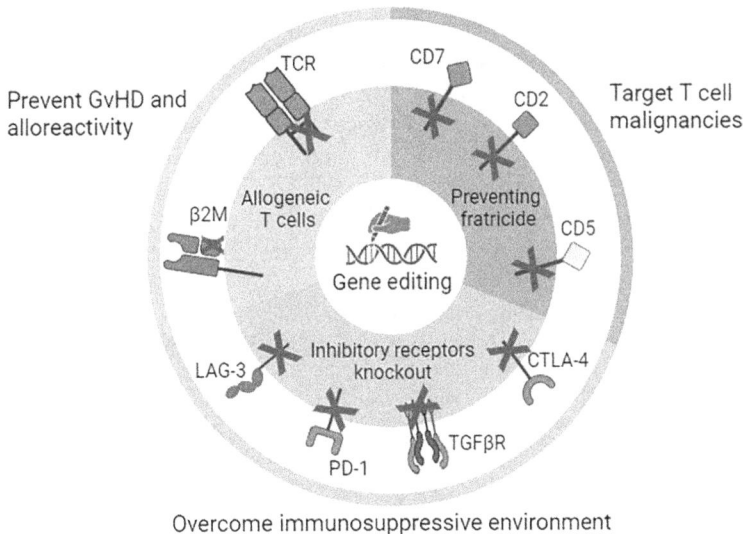

Figure 3.4 Applications of genomic editing in CAR-T cell therapy. Gene editing can be used to manufacture "universal" CAR-T cells by TCR, β2M deletion, target T cell malignancies by preventing fratricide, and disabling inhibitory receptors to overcome immunosuppressive TME. Created with BioRender.com.

allogeneic context.[45] This strategy holds the potential to decrease costs, streamline drug administration, and broaden accessibility to T cell therapies for patients with low T cell counts or serious illnesses.

Application of genome-edited CAR-T cells in clinical trials

Several clinical trials have explored the effectiveness of universally engineered CAR-T cells in treating B cell malignancies. These trials include CAR-T cell therapies targeting CD19 alone (trials NCT02808442 and NCT02746952)[46] and a combinatory targeting of CD19 and CD22 (trial NCT04227015).[47] Published reports have revealed excellent anti-leukemic control and manageable side effects in treated patients, which highlight the potential of gene-edited universal CAR-T cell therapies in treating hematologic malignancies, offering hope for improved outcomes in patients with refractory or relapsed diseases.

Genome editing is a crucial tool in the creation of CAR-T cells that can resist fratricide, a major challenge in treating T cell cancers.[48] Fratricide occurs when CAR-T cells attack both cancerous and healthy T cells, limiting the therapy's effectiveness. To overcome this obstacle, genes like *CD7* or *CD5* can be disabled before introducing CARs targeting these antigens (Figure 3.4). This approach has shown success in clinical trials like NCT04538599[49] and NCT04620655,[50] which focus on treating T cell lymphoma. Published trial results from both studies revealed that 80% of patients achieve CR with an acceptable safety profile. These findings underscore the potential of genome-edited fratricide-resistant CAR-T cells as a promising therapeutic approach for T cell malignancies.

Editing the genome of T cells to knockout inhibitory molecules like PD-1, LAG3, CTLA4, and TGFβR (Figure 3.4) represents a promising strategy to prevent T cell exhaustion and inhibition in the tumour microenvironment (TME).[45] The feasibility of disrupting the aforementioned inhibitory molecules was demonstrated in various preclinical animal experiments. In current CAR-T cell clinical trials, two separate studies have highlighted the feasibility of knocking out the *PDCD1* gene, which encodes PD-1. In phase-1 clinical trial NCT03081715, patients with oesophageal cancer were treated with PD-1 knockout CAR-T cells. The treatment was well tolerated, and the median OS was 127 days, demonstrating the safety and efficacy of PD-1-edited CAR-T cells.[51] Similar safety and efficacy were also reported in NCT02793856, in which patients with non-small cell lung cancer were treated with PD-1 knockout CAR-T cell.[52] In 2021, a first-in-class clinical trial (NCT04976218) was registered for testing TGFβR KO, epidermal growth factor receptor (EGFR) CAR-T cell in patients with EGFR overexpressed tumours. These findings highlight the potential of disabling inhibitory receptors in CAR-T cell via genome editing as a promising strategy to overcome TME suppression in cancer immunotherapy.

RNA transfection

Overview of mRNA engineering in CAR-T cells

Using RNA that codes for proteins to improve the functions and characteristics of immune cells can boost their antitumour efficacy, enhancing their therapeutic potential in cancer treatments. Typically, mRNA is transcribed in vitro and introduced into cells via either electroporation or lipid nanoparticle (LNP) delivery. Clinical studies have shown that introducing CAR encoding mRNA into human T cells via electroporation is feasible and can equip these cells with the ability to destroy tumours.[38,53]

While integrating vectors can ensure that modified CAR-T cells can exert a long-lasting antitumour control due to the stable expression of CAR on the T cells, it also poses the risk of causing "on-target off-tumour toxicity" due to the targeting of tumour-associated antigens that may also be minimally present on healthy tissues.[38] An advantage of using RNA-based transfection is its transient nature, where CAR expression on the T cell surface typically lasts 4–7 days, which can potentially mitigate toxicity effects.[38,53]

Successful translation of in vitro transcribed mRNA into protein is dependent on the translational efficiency and structural stability of the mRNA, which is determined by elements on the mRNA, such as the 5' cap structure, poly(A) tail, coding sequence composition, and untranslated regions (UTRs) present on the molecule's 5' and 3' ends, which can vastly affect the translation efficiency of the mRNA.[54] To bolster stability and increase protein expression levels, modifications can be applied to the mRNA molecule. These modifications include altering the 5' cap structure to protect against degradation by exonucleases, extending the length of the poly(A) tail to improve stability. Additionally, the 5' and 3' UTRs can be modified, thereby enhancing translation efficiency.[34,38,53,54]

Application of mRNA-modified CAR-T cell in the clinics

As listed at www.clinicaltrials.gov, most clinical trials involving mRNA-based CAR-T therapies are phase I trials.[34,38,53] In a recent phase 2 clinical trial (NCT04146051), conducted by Cartesian Therapeutics, 14 patients with myasthenia gravis, an autoimmune disease, were successfully treated with mRNA-engineered anti-BCMA CAR-T therapy.[55] This success highlights the potential of mRNA-engineered CAR-T cells as a novel treatment avenue for autoimmune diseases.

Due to the transient expression of CAR on T cells, a common theme shared across the clinical trials was the requirement for repeated infusion of mRNA-engineered CAR-T cells, which was necessary to ensure sustained tumour eradication While most completed studies reported that treated patients given repeated doses of CAR-T cells did not present treatment-related adverse events (trAEs) such as CRS or ICANs, a significant logistic challenge was reported in clinical trial NCT02623582, where the team was able to manufacture only 60% of the planned CAR-T cell doses.[56] To put it into perspective, in the clinical trial, a single CAR-T dose was administered at 4×10^6/kg, and a total of three doses were required. Assuming a body weight of about 70 kg, a total of $>10^9$ cells are required to be manufactured, accounting for potential cell loss due to electroporation and cryopreservation. While thawing of cryopreserved mRNA-engineered T cells has no impact on CAR expression as demonstrated in rigorous preclinical testing, the need for a high number of cells can be significantly impeded if the patient's T cell quality is compromised and unable to proliferate robustly during the manufacturing process.

Future of RNA transfection in CAR-T cell engineering

While the development of mRNA-modified CAR-T cells is progressing slowly, significant advancements in mRNA engineering have been developed over the past three years.[53] These recent innovations have the potential to generate next generation of mRNAs with greater stability and more robust expression, and could facilitate the application of mRNAs in CAR-T cell therapy, which typically demands higher and sustained mRNA expression compared to mRNA vaccines. Currently, research institutes and biotech companies, such as oRNA Therapeutics, are actively exploring the use of self-amplifying RNA/trans-amplifying RNA (saRNA/taRNA) and circular RNA (circRNA) as alternatives to linear mRNA for genetic engineering of T cells.[57,58] In vaccine applications, saRNA/taRNA and circRNA have demonstrated superior performance to mRNA vaccines, attributed to extended protein expression and stability.[59] Addressing the main concern of transient CAR expression and the need for repeated dosing in mRNA-based CAR-T cells, the use of saRNA and circRNA to deliver the CAR transgene may potentially prolong CAR expression on T cells in vivo, and consequently reduce the number of doses required to achieve therapeutic efficacy and the number of T cells required to be manufactured.

In addition, advancements in mRNA delivery have sparked interest in delivering mRNA encapsulated in lipid nanoparticles (LNPs) to transfect T cells in vivo.[53,57,60] Although there are also ongoing efforts in in vivo transfection of T cells using viral vectors, the transient nature of RNA could offer distinct advantages, potentially minimizing unwanted off-target effects, such as the integration of the transgene into other cell types.

Future perspectives

Throughout this chapter, an overview of the major genetic tools employed for CAR-T cell engineering in clinical trials was presented. While there is ongoing debate about the optimal genetic tool for researchers or biotech companies to adopt, a combination of various genetic tools, with genome editing as a foundational element, is the way forward for achieving precise engineering of T cells. Indeed, with the increasing trend of moving towards the use of an allogeneic T cell source for CAR-T cell therapy in the clinical trials, the current two-pronged approach for the manufacture of allogeneic CAR-T cells is usually mediated by a combination of CRISPR/Cas9 knock-out of *TCR* genes followed by using viral vectors to deliver donor DNA templates for gene knock-in into T cells for the simultaneous deletion of endogenous TCR and insertion of the CAR construct, highlighting the synergy achieved by both genetic tools.[45] Recently, a CRISPR/Cas-associated transposon (CAST) system was discovered, combining the advantages of precision targeting by the CRISPR/Cas protein and the higher efficiency of transgene integration by the transposon. Similarly, Kovač et al. demonstrated the concept of RNA-guided transposition by the SB system,[61] fusing SB transposase with the Cas9 protein, suggesting an attractive strategy for testing in T cells. Finally, to potentially mitigate the risk of insertional mutagenesis, safer non-integrating alternatives, such as RNA-based delivery, can be combined with CRISPR/Cas-edited allogeneic T cells, creating a potential on-demand CAR-T cell therapy. Looking ahead, advancements in current genetic engineering technologies that could enable more accurate and targeted modifications of T cells hold the potential to enhance therapeutic outcomes while minimizing off-target effects and ensuring patient safety.

References

1. Gonzales Carazas, M.M., J.A. Pinto, and F.L. Casado, Biological Bases of Cancer Immunotherapy. *Expert Rev Mol Med*, 2021. **23**: p. e3.
2. Saez-Ibañez, A.R., et al., Landscape of Cancer Cell Therapies: Trends and Real-world Data. *Nat Rev Drug Discov*, 2022. **21**(9): p. 631–632.

3. Rosenberg, S.A., P. Spiess, and R. Lafreniere, A New Approach to the Adoptive Immunotherapy of Cancer with Tumor-infiltrating Lymphocytes. *Science*, 1986. **233**(4770): p. 1318–1321.

4. Topalian, S.L., et al., Immunotherapy of Patients with Advanced Cancer Using Tumor-infiltrating Lymphocytes and Recombinant Interleukin-2: A Pilot Study. *J Clin Oncol*, 1988. **6**(5): p. 839–853.

5. Majzner, R.G. and C.L. Mackall, Clinical Lessons Learned from the First Leg of the CAR T Cell Journey. *Nat Med*, 2019. **25**(9): p. 1341–1355.

6. Zhao, L. and Y.J. Cao, Engineered T Cell Therapy for Cancer in the Clinic. *Front Immunol*, 2019. **10**: p. 2250.

7. Manfredi, F., et al., TCR Redirected T Cells for Cancer Treatment: Achievements, Hurdles, and Goals. *Front Immunol*, 2020. **11**: p. 1689.

8. Sadelain, M., R. Brentjens, and I. Rivière, The Basic Principles of Chimeric Antigen Receptor Design. *Cancer Discov*, 2013. **3**(4): p. 388–398.

9. Liu, Y., et al., TCR-T Immunotherapy: The Challenges and Solutions. *Front Oncol*, 2021. **11**: p. 794183.

10. Holstein, S.A. and M.A. Lunning, CAR T-Cell Therapy in Hematologic Malignancies: A Voyage in Progress. *Clin Pharmacol Ther*, 2020. **107**(1): p. 112–122.

11. Mitra, A., et al., From Bench to Bedside: The History and Progress of CAR T Cell Therapy. *Front Immunol*, 2023. **14**: p. 1188049.

12. Cao, X., et al., China Enters CAR-T Cell Therapy Era. *Innovation (Camb)*, 2022. **3**(1): p. 100197.

13. Gardner, R.A., et al., Intent-to-treat Leukemia Remission by CD19 CAR T Cells of Defined Formulation and Dose in Children and Young Adults. *Blood*, 2017. **129**(25): p. 3322–3331.

14. Neelapu, S.S., et al., Axicabtagene Ciloleucel CAR T-Cell Therapy in Refractory Large B-Cell Lymphoma. *N Engl J Med*, 2017. **377**(26): p. 2531–2544.

15. Maude, S.L., et al., Tisagenlecleucel in Children and Young Adults with B-Cell Lymphoblastic Leukemia. *N Engl J Med*, 2018. **378**(5): p. 439–448.

16. Schuster, S.J., et al., Tisagenlecleucel in Adult Relapsed or Refractory Diffuse Large B-Cell Lymphoma. *N Engl J Med*, 2019. **380**(1): p. 45–56.

17. Hopewell, E.L., C. Cox, S. Pilon-Thomas, and L.L. Kelley, Tumor-infiltrating Lymphocytes: Streamlining a Complex Manufacturing Process. *Cytotherapy*, 2019. **21**(3): p. 307–314.

18. Levine, B.L., J. Miskin, K. Wonnacott, and C. Keir, Global Manufacturing of CAR T Cell Therapy. *Mol Ther Methods Clin Dev*, 2017. **4**: p. 92–101.

19. Milone, M.C. and U. O'Doherty, Clinical Use of Lentiviral Vectors. *Leukemia*, 2018. **32**(7): p. 1529–1541.

20. Maier, P., C. von Kalle, and S. Laufs, Retroviral Vectors for Gene Therapy. *Future Microbiol*, 2010. **5**(10): p. 1507–1523.

21. Cronin, J., X.Y. Zhang, and J. Reiser, Altering the Tropism of Lentiviral Vectors through Pseudotyping. *Curr Gene Ther*, 2005. **5**(4): p. 387–398.

22. Gutierrez-Guerrero, A., F.L. Cosset, and E. Verhoeyen, Lentiviral Vector Pseudotypes: Precious Tools to Improve Gene Modification of Hematopoietic Cells for Research and Gene Therapy. *Viruses*, 2020. **12**(9).

23. Nikolic, J., et al., Structural Basis for the Recognition of LDL-receptor Family Members by VSV Glycoprotein. *Nat Commun*, 2018. **9**(1): p. 1029.

24. Amirache, F., et al., Mystery solved: VSV-G-LVs Do Not Allow Efficient Gene Transfer Into Unstimulated T Cells, B Cells, and HSCs Because They Lack the LDL Receptor. *Blood*, 2014. **123**(9): p. 1422–1424.

25. Marcucci, K.T., et al., Retroviral and Lentiviral Safety Analysis of Gene-Modified T Cell Products and Infused HIV and Oncology Patients. *Mol Ther*, 2018. **26**(1): p. 269–279.

26. FDA U.S.A, Considerations for Development-of-Chimeric Antigen Receptor T Cell Products. 2024.

27. Mhaidly, R. and E. Verhoeyen, The Future: In vivo CAR T Cell Gene Therapy. *Mol Ther*, 2019. **27**(4): p. 707–709.

28. Kathryn, R.M., et al., Preclinical Proof of Concept for VivoVec, a Lentiviral-based Platform for in Vivo CAR T-cell Engineering. *J Immunother Cancer*, 2023. **11**(3): p. e006292.

29. Vigant, F., et al., Abstract 3294: In Vivo Delivery of a Novel CD3-targeted Lentiviral Vector Generates CD19 CAR-T Cells in Two Different Humanized Mouse Models and Results in Complete B Cell Depletion. *Cancer Res*, 2022. **82**(12_Supplement): p. 3294–3294.

30. Andorko, J.I., et al., Targeted In vivo Generation of CAR T and NK Cells Utilizing an Engineered Lentiviral Vector Platform. *Blood*, 2023. **142**(Supplement 1): p. 763–763.

31. Cliff, E.R.S., et al., High Cost of Chimeric Antigen Receptor T-Cells: Challenges and Solutions. *Am Soc Clin Oncol Educ Book*, 2023(43): p. e397912.

32. Levine, B.L., et al., Unanswered Questions Following Reports of Secondary Malignancies After CAR-T Cell Therapy. *Nat Med*, 2024. **30**(2): p. 338–341.

33. Sandoval-Villegas, N., W. Nurieva, M. Amberger, and Z. Ivics, Contemporary Transposon Tools: A Review and Guide through Mechanisms and Applications of Sleeping Beauty, piggy-Bac and Tol2 for Genome Engineering. *Int J Mol Sci*, 2021. **22**(10).

34. Moretti, A., et al., The Past, Present, and Future of Non-Viral CAR T Cells. *Front Immunol*, 2022. **13**: p. 867013.

35. DeNicola, G.M., F.A. Karreth, D.J. Adams, and C.C. Wong, The Utility of Transposon Mutagenesis for Cancer Studies in the Era of Genome Editing. *Genome Biol*, 2015. **16**: p. 229.

36. Gogol-Döring, A., et al., Genome-wide Profiling Reveals Remarkable Parallels Between Insertion Site Selection Properties of the MLV Retrovirus and the piggyBac Transposon in Primary Human CD4(+) T Cells. *Mol Ther*, 2016. **24**(3): p. 592–606.

37. Kebriaei, P., et al., Phase I Trials Using Sleeping Beauty to Generate CD19-Specific CAR T Cells. *J Clin Invest*, 2016. **126**(9): p. 3363–3376.

38. Irving, M., et al., Choosing the Right Tool for Genetic Engineering: Clinical Lessons from Chimeric Antigen Receptor-T Cells. *Hum Gene Ther*, 2021. **32**(19–20): p. 1044–1058.

39. Bishop, D.C., et al., Development of CAR T-cell Lymphoma in 2 of 10 Patients Effectively Treated with PiggyBac-Modified CD19 CAR T Cells. *Blood*, 2021. **138**(16): p. 1504–1509.

40. Micklethwaite, K.P., et al., Investigation of Product-Derived Lymphoma Following Infusion of PiggyBac-modified CD19 Chimeric Antigen Receptor T Cells. *Blood*, 2021. **138**(16): p. 1391–1405.

41. Gregory, T., et al., Efficacy and Safety of P-Bcma-101 CAR-T Cells in Patients with Relapsed/ Refractory (r/r) Multiple Myeloma (MM). *Blood*, 2018. **132**(Supplement 1): p. 1012–1012.

42. Costello, C.L., et al., Phase 2 Study of the Response and Safety of P-Bcma-101 CAR-T Cells in Patients with Relapsed/Refractory (r/r) Multiple Myeloma (MM) (PRIME). *Blood*, 2019. **134**(Supplement_1): p. 3184–3184.

43. Shamshirgaran, Y., et al., Tools for Efficient Genome Editing; ZFN, TALEN, and CRISPR, in *Applications of Genome Modulation and Editing*, P.J. Verma, H. Sumer, and J. Liu, Editors. 2022, Springer US: New York, NY. p. 29–46.

44. Depil, S., et al., 'Off-the-shelf' Allogeneic CAR T Cells: Development and Challenges. *Nat Rev Drug Discov*, 2020. **19**(3): p. 185–199.

45. Dimitri, A., F. Herbst, and J.A. Fraietta, Engineering the Next-generation of CAR T-cells with CRISPR-Cas9 Gene Editing. *Mol Cancer*, 2022. **21**(1): p. 78.

46. Benjamin, R., et al., Genome-edited, Donor-Derived Allogeneic anti-CD19 Chimeric Antigen Receptor T Cells in Paediatric and Adult B-cell Acute Lymphoblastic Leukaemia: Results of Two Phase 1 Studies. *Lancet*, 2020. **396**(10266): p. 1885–1894.

47. Hu, Y., et al., CRISPR/Cas9-Engineered Universal CD19/CD22 Dual-targeted CAR-T Cell Therapy for Relapsed/Refractory B-cell Acute Lymphoblastic Leukemia. *Clin Cancer Res*, 2021. **27**(10): p. 2764–2772.

48. Luo, L., et al., Current State of CAR-T Therapy for T-cell Malignancies. *Ther Adv Hematol*, 2022. **13**: p. 20406207221143025.

49. Hu, Y., et al., Genetically Modified CD7-Targeting Allogeneic CAR-T cell Therapy with Enhanced Efficacy for Relapsed/Refractory CD7-positive Hematological Malignancies: A Phase I Clinical Study. *Cell Res*, 2022. **32**(11): p. 995–1007.

50. Zhang, X., et al., A Novel Universal CD7-Targeted CAR-T Cell Therapy for Relapsed Or Refractory T-cell Acute Lymphoblastic Leukemia and T-cell Lymphoblastic Lymphoma. *Blood*, 2022. **140**(Supplement 1): p. 4566–4567.

51. Jing, Z., et al., Safety and Activity of Programmed Cell Death-1 Gene Knockout Engineered T Cells in Patients with Previously Treated Advanced Esophageal Squamous Cell Carcinoma: An Open-label, Single-arm Phase I Study. *J Clin Oncol*, 2018. **36**(15_suppl): p. 3054–3054.

52. Liu, Q., World-First Phase I Clinical Trial for CRISPR-Cas9 PD-1-Edited T-Cells in Advanced Nonsmall Cell Lung Cancer. *Glob Med Genet*, 2020. **7**(3): p. 73–74.

53. Xiao, K., et al., mRNA-based Chimeric Antigen Receptor T Cell Therapy: Basic Principles, Recent Advances and Future Directions. *Interdisc Med*, 2024. **2**(1): p. e20230036.
54. Kang, D.D., H. Li, and Y. Dong, Advancements of in Vitro Transcribed mRNA (IVT mRNA) to Enable Translation into the Clinics. *Adv Drug Deliv Rev*, 2023. **199**: p. 114961.
55. Granit, V., et al., Safety and Clinical Activity of Autologous RNA Chimeric Antigen Receptor T-cell Therapy in Myasthenia Gravis (MG-001): A Prospective, Multicentre, Open-label, Non-randomised Phase 1b/2a Study. *Lancet Neurol*, 2023. **22**(7): p. 578–590.
56. Cummins, K.D., et al., Treating Relapsed / Refractory (RR) AML with Biodegradable Anti-CD123 CAR Modified T Cells. *Blood*, 2017. **130**: p. 1359.
57. Mabry, R., et al., 1222 In situ CAR Therapy Using oRNA™ Lipid Nanoparticles Regresses Tumors in Mice. *J Immunother Cancer*, 2022. **10**(Suppl 2): p. A1267–A1267.
58. Shen, L., et al., Circular mRNA-based TCR-T offers a Safe and Effective Therapeutic Strategy for Treatment of Cytomegalovirus Infection. *Mol Ther*, 2024. **32**(1): p. 168–184.
59. Rohner, E., et al., Unlocking the Promise of mRNA Therapeutics. *Nat Biotechnol*, 2022. **40**(11): p. 1586–1600.
60. Billingsley, M.M., et al., In vivo mRNA CAR T Cell Engineering via Targeted Ionizable Lipid Nanoparticles with Extrahepatic Tropism. *Small*, 2024. **20**(11): p. e2304378.
61. Kovač, A., et al., RNA-guided Retargeting of Sleeping Beauty Transposition in Human Cells. *elife*, 2020. **9**.

4 Intracellular delivery methods for therapeutic immune T-cell engineering

Arun R. K. Kumar and Andy Kah Ping Tay

Introduction

Cancer is a complex multifactorial disease that leads to 19.3 million new cases and about 10 million deaths every year.[1] Although traditional treatments such as surgery, chemotherapy, and radiotherapy can be successful in early stages of the tumor, treatment is still challenging for cases of relapsed/refractory diseases. Adoptive T cell immunotherapy has emerged as a forefront in treating patients with advanced cancers by genetically modifying the patient's own immune cells to target the tumor, thus offering a therapeutic advantage over conventional methods. Tumor-infiltrating lymphocyte (TIL) T cell therapy is a form of adoptive therapy that relies on selectively finding tumor-reactive lymphocytes in the excised tumor from patients, which are then expanded in large numbers ex vivo for patient reinfusion. Although there was initial success from TIL therapies in cervical carcinoma[2] and metastatic melanoma, it was later reported that TILs have low persistence in vivo.[3,4] Other adoptive approaches being developed include T cell receptor (TCR)–based cell therapy and chimeric antigen receptor (CAR)-T cell therapy, both of which function by genetically modifying peripheral blood-derived lymphocytes with TCR subunits or CAR receptors to elicit a tumor-specific antigen response, respectively. TCR-T cells are engineered by introducing modified TCR α and β chains into T cells that have high affinity to a specific tumor-expressing antigen. This approach exploits the interactions of the newly adopted TCR with the Major Histocompatibility Complex (MHC) complex present on the tumor cell surface, which subsequently activates the T cells for antitumor activity.[5] For an effective TCR-T response, the target antigen should be selectively overexpressed on tumors, and their expression on benign tissues must be very low. Yet, TCR-T cell therapies are ridden with drawbacks. The introduction of the novel αβ chain has been associated with risks of TCR chain mispairing with the endogenous TCR chains.[6] This leads to self-reactive TCR clones and off-target toxicity, rendering the TCR-T cells ineffective. TCR-T cells lack costimulatory domains and additionally because TCR-T cells are dependent on MHC presentation, they have reduced persistence in the tumor.[5,7] In addition, the limited TCR repertoire combined with concerns looming over the target affinity and cross-reactivity poses challenges for large-scale adoptive cell therapeutics using TCR-T cells.[8,9]

Advent of the CAR-T era

CAR-T cell therapy relies on genetically modifying CD3+ T cells to express a tumor targeting receptor or CAR for tumor elimination in patients. Unlike TCRs, CARs consist of

DOI: 10.1201/9781032660752-4

an external single-chain variable fragment antigen-binding domain and an intracellular CD3ζ signaling domain that help in T cell activation. This eliminates the need for MHC presentation and allows CARs to bind and attack virtually any cell-surface antigen expressed by the tumor cells. Subsequent research demonstrated that the long-term survival and cytotoxic activity of CAR-T cells could be enhanced by incorporating one or more costimulatory signaling domains[10,11] (CD28, 4-1BB, or OX40) and an inducible cytokine expression transgene[12] (interleukin-12 [IL-12] and interleukin-15 [IL-15]) into the structure of the CAR receptor.

Following many pioneering preclinical studies in B-cell malignancies, the first clinical application for engineered CAR-T cells was reported in 2010 when CD19-targeting CAR-T cells showed regression of B-cell leukemia.[13] Clinical trials involving engineered CD19 CAR-T cells showed notable success in inducing complete remissions in patients with relapsed/refractory B-cell lymphoma, B-cell leukemia, and mantle cell lymphomas.[14-16] As a result, this success has led to the approval of six CAR-T therapies for clinical use by the U.S. Food and Drug Administration (FDA) till date: Kymriah from Novartis Yescarta and Tecartus from Kite Pharma/Gilead Sciences, Breyanzi and Abecma from Bristol-Myers Squibb, and Carvykti from Janssen Biotech.[17] Beyond the United States, the CAR-T products have received marketing authorization from the European Medicines Agency for clinical use in Europe. And as of May 2024, there are at least 1500 clinical trials focused on CAR-T therapies listed on ClinicalTrials.gov showing continued interest in using genetically engineered T cell therapies for cancer treatments.

CAR engineering beyond CAR-T cells

Apart from T cells, other immune effector cells such as natural killer (NK) cells, gammadelta (γδ)-T cells, dendritic cells (DCs), and macrophages have emerged as alternatives for T cell-based CAR therapy. CAR-NK cells provide innate immunity against tumor cells, and due to the absence of a TCR, they have minimal potential to induce GvHD (graft versus host disease). This allows clinicians to consider CAR-NKs as an "off-the-shelf" allogeneic product for clinical use to treat multiple patients.[18] Besides, genetic engineering of NK cells with an additional CAR receptor improves the antitumor response and cytokine production of interleukin-3, interferon-gamma (IFN-γ), and granulocyte-macrophage colony stimulating factor.[19,20] Anti-CD19 CAR-NK cells were able to offer complete response in 64% patients presented with non-Hodgkin's lymphoma and chronic lymphocytic leukemia in a phase I/II trial.[21] In spite of the advances, poor in vivo expansion, short-term persistence, and trafficking into solid tumors suggest that CAR-NK[22] in large-scale therapeutics might be challenging. At the same time, applicability and efficacy of CAR-NK for solid tumors are yet to be validated. γδ-T cells form a very small subset of T cells of about 2% in the peripheral blood T cells but are found in larger proportions in gut mucosa.[23] They differ from T cells, which express the conventional αβ TCR chain, thereby offering MHC-independent antitumor immune activity. γδ-T cells serve as an appealing alternative to conventional αβ CAR-T cells, particularly when the latter are unsuitable for patients due to cytokine-associated toxicity and on-target, off-tumor toxicity.[24] However, it should be noted that only a handful of clinical trials are currently being conducted using CAR-γδ T cells.

Regulatory T cells are a subset of CD4+ T cells identified by their characteristic overexpression of CD25 and FoxP3 and well known for their immunosuppressive function in the immune system. CAR-T regulatory (CAR-T$_{reg}$) cells on the contrary are another promising alternative to CAR-T critical in maintaining T cell immune tolerance while effectuating an antitumor activity.[25] Immune tolerance after CAR-T$_{reg}$ injection would reduce the chances of graft rejection and is heavily being investigated for autoimmune conditions, notably colitis,[26] asthma,[27] and vitiligo.[28] CAR-T$_{reg}$ cells are also in the investigative phase of human trials for maintaining immunological tolerance after organ transplants (NCT04817774, NCT05234190).[29,30] Besides, CAR-modified T$_{reg}$ cells are less dependent on interleukin-2 (IL-2) for proliferative signals and exhibit antigen specificity, in contrast to polyclonal TCR T$_{reg}$ cells. Research into genetic engineering of CAR-DCs is underway, and recent studies have demonstrated that CAR-DCs can enhance T cell cytotoxicity when combined with CAR-T cells. When CAR-DCs are combined with CAR-T cells, there is a notable increase in cytotoxicity and higher cytokine expression levels of IFN-γ, IL-2, and tumor necrosis factor-**α** (TNF-**α**) compared to treatment with CAR-T cells alone.[31] But these promising preclinical results need further investigation and eventual clinical trials to assess the safety and efficacy of CAR-DCs in large-scale therapeutics. Macrophages are tissue-resident immune cells that when activated can mediate phagocytosis of pathogens and dead cells, induce inflammatory response, and maintain T cell tolerance in healthy tissues.[32] Recent advancements in macrophage-based cell therapy have focused on differentiating patient-derived monocytes into macrophages and subsequently transducing or transfecting them to express a CAR receptor. These engineered macrophages are then primed to infiltrate solid tumors[33] and induce pro-inflammatory cytotoxic responses toward tumors in vivo.[34] CAR-macrophage (CAR-M) cells infiltrating the tumor environment can thereafter recruit both activated and resting T cells to the tumor site. CAR-M is also shown to promote the maturation of DCs and induce a pro-inflammatory phenotype in M2 macrophages.[34]

Clinical CAR-T manufacturing challenges

The production of CAR-T cells involves several meticulously executed steps, and before they are cleared for patient infusion, continuous quality control testing is performed to assess the safety profiles and potency of the final CAR-T product[35] (Figure 4.1). While protocols for manufacturing clinical-grade CAR-T cells have been established, the number of patients benefiting from CAR-T cell therapies appears to be in the low tens of thousands so far. Due to the personalized approach of current CAR-T therapy, there is a substantial cost of manufacturing for a single CAR-T dose. Recent estimates suggest that the U.S. FDA–approved Kymriah (Novartis) costs about US$475,000, excluding the clinic-associated costs.[36] Addressing challenges associated with large-scale manufacturing costs and long-term safety concerns arising from the treatment will be essential to advancing CAR-T cell therapies and making them more widely accessible to patients with diverse cancer types. Despite clinical success, several challenges remain in the efficacy, safety, and manufacture of CAR-T cells, which can impair CAR-T cell functionalities and ultimately the outcome of therapy. Patients receiving CAR-T therapy often suffer from treatment-related toxicities and severe adverse side effects, which can even lead to death.[37–40]

Autologous T cells are primarily utilized in CAR-T cell manufacturing of majority of clinical trials and U.S. FDA–approved therapies. This approach aims to limit graft rejection, rendering CAR-T therapy personalized. However, in certain patient demographics such as those presented with advanced cancers, older individuals, or pediatric patients, there could be an insufficient number of circulating healthy naive T cells in peripheral blood. This scarcity can be attributed to several factors such as prior lines of drug

b

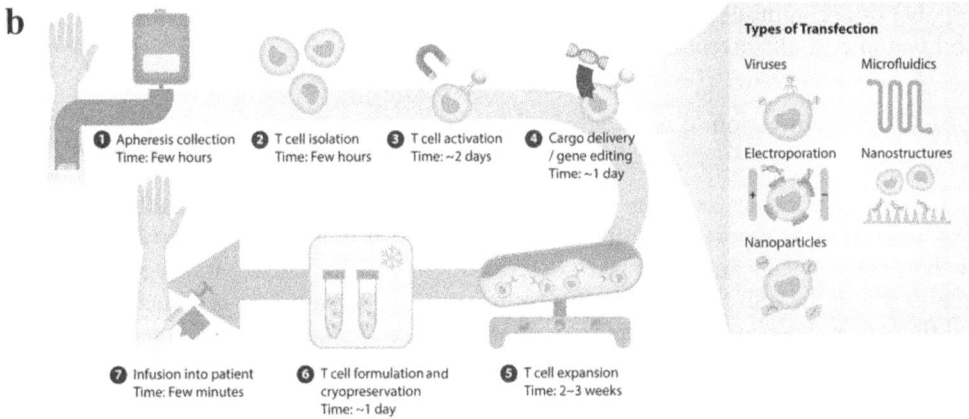

Figure 4.1 CAR-immune cell manufacturing workflow and transfection. The CAR-T cell manufacturing process begins with the extraction of blood from the patient, followed by the isolation of the targeted T cell type and then stimulation and activation of T cells. Subsequently, the T cells are transfected with CAR transgenes. Researchers can choose between different transfection tools that are available for genome engineering of primary T cells such as viral vectors and bulk electroporation and emerging tools such as microfluidics, nanoparticles, and high-aspect-ratio nanostructures. The transfected cells are expanded in sufficient numbers, cryopreserved, and shipped to clinics for patient reinfusion. Reproduced with permission.[43] Copyright 2021, Wiley-VCH GmbH.

regimens, therapies, or age. Consequently, the limited number of healthy naive T cells significantly impacts the manufacturing process, potentially hindering the expansion of sufficient cells for a single CAR-T infusion. CAR-T manufacturing is an intricate process that necessitates the expansion of transfected CAR-T cells in sufficient numbers. Additionally, each CAR-T product typically needs 10^6–10^7 CAR-T cells/kg of patient. However, due to the aforementioned reasons, patient-derived T cells may inadequately divide, resulting in substandard products that fail to function effectively at the tumor site.[41,42] These manufacturing failures can be critically linked to the quality of cells derived from patients and the complexity of CAR constructs delivered into the T cells. Allogeneic CAR-T cells present a potential solution to these manufacturing challenges, enabling the advent of so-called off-the-shelf CAR-T products. These products can be utilized for multiple patients, eliminating hindrances associated with poor immune cell quality from individual patients.

Additionally, the CAR-T cell manufacturing process typically lasts for about two weeks. However, logistical challenges such as shipping apheresis products and meeting the release criteria, including transduction efficiency and cell viability, can extend the overall process from vein to vein from enrollment to infusion beyond two weeks. A recent review revealed that about 7% of patients[44] did not survive while awaiting the completion of CAR-T cell manufacturing, emphasizing the urgency in shortening the production times.[45] It is currently evident that progress in CAR-T therapeutics needs pioneering solutions, strategies beyond targeting tumors using a single antigen-specific target such as checkpoint blockage, combination of cytokine secretion, and better immune trafficking into solid tumors. As these strategies typically require complex CAR designs, a need for engineering strategies to deliver them into immune cells becomes indispensable. Achieving this requires highly efficient and safe transfection techniques capable of accommodating various payloads and CAR construct sizes, along with scalability and cost-effectiveness.[46,47]

CAR-T cell transfection

A primary challenge in CAR-T production is effectively delivering the CAR transgene into hard-to-transfect primary T cells and expanding the CAR-T cells ex vivo.[48] Transfection is a critical technique that helps in delivery of exogenous biological cargo such as nucleic acids and proteins to functionally manipulate genes and regulate protein expression in the cells.[49] However, this technique is hindered by the cell membrane (typical thickness of 5–6 nm) that serves as an essential barrier to exogenous molecules.

Achieving stable transfection involves integrating the transgene into the host genome, ensuring prolonged expression even through generations of cell division. Stable expression vectors are often preferred for CAR-T cell generation to ensure prolonged CAR expression on T cells.[50] Viral vectors, CRISPR knock-ins, or transposons typically achieve stable gene transfection, while intracellular delivery of mRNA, small interfering RNA (siRNA), DNA plasmids, or oligonucleotides leads to transient expression due to non-integration with the genome.[51] In gene and cell therapies, transfection plays a pivotal role in efficiently delivering transgenes to target cells and tissues, aiding in gene-corrective therapies or producing genetically modified clones for long-term therapeutic effect in patients. Despite its significance, transfection remains a bottleneck due to limitations of current technologies, prompting development of next-generation solutions in intracellular delivery.[49]

T cell genome engineering modalities

Viral transduction

Transfection using viral vectors or viral transduction capitalizes on the innate ability of viruses to infect various host mammalian cell types and inject genetic material, with the transfer plasmid encoding the transgene. Using several viral enzymes such as reverse transcriptase and integrase, the transfer cassette integrates into the host genome, subsequently resulting in long-term CAR transgene expression in host T cell clones. Viral vectors are amenable to both in vivo and in vitro genetic medication of T cells.

Considerable costs and resources are devoted to designing viral vectors to eliminate their infectivity and replicability in the host cells. This is ensured by replacing the viral sequences necessary for replication and packaging with the transgene coding sequences. They are subjected to rigorous testing for competency, toxicity, and neoplastic potential. It's noteworthy that all U.S. FDA–approved CAR-T therapies to date utilize viruses for T cell genetic modification. Yescarta and Tecartus employ gamma-retroviral vectors, while others utilize lentiviruses (LVs). Currently, clinical manufacturing of CAR-T cells heavily relies on viral vectors due to their high transduction rates and sustained expression of transgenes. Retroviral vectors can transduce only dividing cells, whereas lentiviral vectors can infect both dividing and non-dividing cells, albeit T cell activation tends to promote gene transduction efficacies. Vormittag et al. reported that approximately 94% of clinical-grade T cells are manufactured using viral transduction machinery, with 54% of CAR-T products utilizing LVs and the remainder employing retroviruses (RVs).[52]

While viruses are widely regarded as the gold standard in the transfection industry, they face numerous biological and manufacturing challenges in cell therapy manufacturing. RVs, for instance, prompt genomic integration near transcriptional start sites and CpG islands. Such integrations around promoters and enhancers elevate the risk of oncogenic transformation by activating proto-oncogenes.[53] Follow-up investigations raise concerns regarding insertional mutagenesis and potential for further malignant development of T cell leukemia in patients with X-linked severe combined immunodeficiency following RV vector integration and LIM only protein 2 (LMO2) gene activation.[54] LVs, on the contrary, have a proclivity to integrate into transcriptionally active sites, which is considered a safe integration mechanism. But a recent study in a CAR-T patient reported that LV integration into the *TET2* gene may impact T cell proliferation and clinical responses; however, further research is needed to fully understand the viral integration machinery in gene therapies.[55,56] Additionally, adeno-associated viruses, which are considered to be usually non-integrating, are also recently reported to cause insertional mutagenesis in preclinical studies.[57,58]

Additionally, viral-based therapies induce transduction-associated genotoxicity and immunogenicity in the host cells.[59] Viral-based gene therapies cause adverse immune reaction in patients, both humoral and adaptive response against viral vector-derived immunogenic epitopes and transgene.[60] Viral-induced immunogenicity reduces the efficacy of the therapy, limits T cell functionality, and exacerbates the incidence of Cytokine Release Syndrome (CRS) and CAR-T-associated toxicities. Cromer et al. observed that transduction induced notable changes in the expression of genes associated with pro-inflammatory, metabolic, and cell-cycle-related pathways in primary stem cells.[61] While no fatality has been associated with the use of viral transduction for CAR-T, there is an inherent risk of immunogenicity, abrupt cytokine production, and altered gene expression linked to the use of viral vectors compared to other non-viral transfection methods available.

Following numerous reports of secondary T cell malignancies in patients receiving commercial autologous CAR-T therapies, the U.S. FDA has launched investigations into the serious risk of T cell malignancy associated with these treatments.[62] Cases of CAR-positive T cell lymphoma have been reported in patients as early as three months after the treatment.[63] Due to the integration of lentiviral or retroviral vectors used in these therapies, U.S. FDA has determined that there is a recognized risk of developing secondary malignancies labeled as a class warning in the U.S. prescribing information. U.S. FDA recommendations also emphasize on lifelong monitoring of patients and clinical trial participants who receive these therapies for the development of new malignancies, underscoring the importance of continuous vigilance in using CAR-T products engineered with integrating viruses. Consequently, the initial approvals of these products included mandates conducting long-term follow-up observational safety studies spanning 15 years to assess the safety and efficacy posttreatment.[62,64]

In addition to the previously mentioned challenges, viral vectors have cargo size limitations, typically ranging from 8 to 9 kb for most vectors,[65] extending up to 15 kb in certain adenoviruses.[48,66] As CAR-construct designs evolve to accommodate additional costimulatory signal domains, cytokine-expressing cassettes, gene-editing elements encoding Cas9 expression, and oligonucleotides like siRNA or gRNA, the practicality of incorporating them into a single-vector backbone with limited packaging capacity becomes increasingly challenging.[11,12,67] While high transgene expression is crucial to improve the utility of CAR-T, attempting to include complex CAR design cassettes may require the cotransduction of two or more vectors. However, successfully transducing multiple vectors into the same cell has proven to be inefficient, impacting the transduction efficiency negatively. Moreover, viral transduction is limited to introducing nucleic acids and cannot deliver other relevant biomolecules such as oligonucleotides, peptides, or ribonucleoproteins (RNPs). Our previous work has demonstrated that viral transduction induces heightened immune responses and cell death compared to transfection techniques like bulk electroporation, lipofectamine, and high-aspect-ratio nanostructures.[68] Moreover, viral transduction triggers an abrupt increase in intracellular calcium levels, indicating T cell stress post-transduction.[68]

Another critical caveat lies in the manufacturing process of viral-based transfection. The U.S. FDA categorizes viral particles as raw materials for clinical cell therapy manufacturing and mandates the production to follow current Good Manufacturing Practices (cGMP) with biosafety level 2 facilities and highly trained personnel. Vector production involves several quality checkpoints to ensure product quality and safety, including the removal of replication-competent viruses at various production stages. U.S. FDA recommendations also include safety monitoring and long-term observational studies of patients for up to 15 years post-infusion to detect any unanticipated infections or vector replication emergence due to treatment. The complex manufacturing processes, rigorous testing for competency, purity, and toxicity, coupled with the need for long-term safety monitoring, result in increased costs and logistical challenges. Consequently, the use of viral vectors may contribute to an expensive final CAR-T product, significantly limiting patient access to CAR-T cell therapy.[69]

Bulk electroporation

To address the manufacturing and complex biological challenges arising from viral vectors, a technique popularized in the 1980s called bulk electroporation uses high-voltage electric pulses (in the tens of kilovolts range) to transiently perforate cell membrane and deliver exogenous cargo such as drugs, molecular beacons, and DNA plasmids into the cells.[70] Cell membranes tend to maintain an electrical potential of ~0.07 V across the lipid–bilayer, which separates the cytosolic ionic difference from the external microenvironment. Bulk electroporation uses high-voltage pulses, ranging from tens to hundreds of volts, which change the membrane potential beyond the critical threshold (~1V) and disturb the membrane arrangement ultimately causing it to break down. This breakdown can be either temporary or irreversible. Subsequently, ion leakage from the cells and the entry of biomolecule cargo from the external microenvironment occur through passive diffusion. In a conventional bulk electroporation setup, cells in suspension are mixed with a suitable conductive buffer solution and placed between two parallel electrode plates within a cuvette, and an electrical generator is connected to produce a voltage difference between the plates.

Bulk electroporation has become the most commonly used non-viral transfection system, and the technology has brought about the commercially available electroporation devices like CliniMACS® Electroporator (Miltenyi Biotec), 4D-Nucleofector® System (Lonza), Gene Pulser XCell™ Electroporation Systems (Bio-Rad), and Neon Transfection System® (Thermo Fisher Scientific). Several of them have even been integrated into fully automated, closed, GMP compliant systems such as the CliniMACS Prodigy (Miltenyi Biotec) and Cocoon (Octane Biotech) for CAR-T cell manufacturing. Such commercial devices come with preset programs and parameters such as electric voltage, frequency, pulse duration, and waveforms optimized for different cell types, both immortalized cell lines and primary. Despite this feature, electroporation into sensitive cell types is still challenging because of significant cell damage and toxicity. Because these programs are not manually modifiable, users are faced with challenges in optimization for different cell types.

Unlike viruses, bulk electroporation is cargo independent; it can deliver any cargo molecule ranging from nucleic acids, oligonucleotides, proteins, fluorescent dyes, and drugs. In literature, electroporation has heavily been used for delivering nucleic acids such as plasmids and mRNA for transient CAR expression on T cells. Due to the transiency of CAR protein expression, engineered CAR-T via mRNA transfection eliminates caveats of genomic integration like safety concerns and adverse immune reactions in the patients. But mRNA transfection also signifies that regular infusions might be needed for long-term CAR-T surveillance in the patients. Almåsbak et al. reported that they achieved 94% CD19 CAR expression in T cells where more than 80% T cells were viable after CAR mRNA electroporation with a square-wave pulse.[71] Impressively, another study reports that more than 80% CD4+ T and CD8+ T cells were bulk electroporated with CAR mRNA encoding anti-ErbB2 and anti-CEA receptors.[72] After transfection, the CAR-T cells exhibited impressive antitumor response against tumors positive for both ErbB2 and CEA receptors, respectively. Their team also reports that there was negligible CAR expression beyond eight days, thereby minimizing the potential risk of auto-aggressive T cell responses.[72]

One of the seminal works with CRISPR/Cas9 gene editing was reported by Schumann et al., who bulk electroporated gene-editing elements such as Cas9 RNPs designed to knock in homology-directed repair donor templates at CXCR4 and PD-1 gene loci of primary T cells. They report an efficiency of ~25% knock-in at the CXCR4 locus and ~20% knock-in at the PD-1 locus.[73] For the first time, Eyquem et al. showed that CAR-T cell electroporation of Cas9 mRNA and gRNA produced about 70% knockout efficiency of the endogenous TCR receptor in human peripheral blood T cells and in turn enhanced T cell potency in vivo against a mouse model of acute lymphoblastic leukemia.[74] Following this, other researchers have also shown the use of bulk electroporation to engineer T cells by delivering the CRISPR/Cas9 editing system to replace the endogenous TCR gene loci for reinvigorating the antitumor response of T cells in vitro and in vivo.[75] A recent research published by Stenger et al. shows that engineered anti-CD19-CAR-T cells bulk electroporated with Cas9 RNPs targeting the B chain of endogenous TCR receptor lose the TCR function in 78.2% of T cell population.[76]

Recent advancements have introduced the use of transposon/transposase plasmid vector systems, such as piggyBac and Sleeping Beauty, to achieve stable expression of transgenes in T cells using bulk electroporation. These non-viral transposition vector systems are characterized by their simplicity, versatility, low immunogenicity, and ease of large-scale production. Moreover, bulk electroporation can be readily employed to transfect transposon/transposase vectors and generate CAR-T cells.[77,78] Early results from phase I/II clinical trials employing bulk electroporation to engineer transposon-based CAR-T cells have demonstrated safety and minimal adverse effects in the long term (trial identifiers: NCT00968760, NCT01497184, and NCT01492036).[79,80] Further discussion on transposon-based CAR-T cell engineering would be followed in the next chapter.

Despite these advantages, bulk electroporation induces significant biological perturbations in cells. The high-voltage pulses cause severe cell membrane shearing and lowers cell viability in primary and sensitive cell types.[81] Studies have shown that bulk electroporation induces significant transfection-induced biological cell stress, as evidenced by increased calcium influx, lower cell viability, altered gene expression profiles, and lengthened cell-doubling times in Jurkat cells (a cancerous T cell line model) compared to other transfection methods.[68] Prolonged cell-doubling times in CAR-T manufacturing can lead to manufacturing delays and impact overall production costs.[68] Additionally, bulk electroporation has been associated with aberrant cytokine expression and altered gene expression in primary immune T cells, resulting in poor efficacy in preclinical experiments conducted in mice. Several of the downsides of bulk electroporation could potentially arise from the high-voltage pulses (tens and hundreds of kV/cm) used in the cuvette chamber. Strong, non-homogenous electric fields have previously been shown to induce Joule heating[70] and an increase in the chamber temperature,[82] adversely impacting the cellular viability. Recent experiments have integrated microfluidic designs into electroporation chambers to minimize the required electric field threshold and mitigate issues such as cell viability, Joule heating, and biological perturbations to transfected cells. These advancements aim to improve the efficiency and safety of bulk electroporation in the context of CAR-T cell manufacturing.[83]

Transfection with microfluidics

Microfluidics-assisted transfection uses microscale fluidic flow[84] to precisely control the permeabilization of cell membrane and deliver exogenous cargo into the cells.[85] Typically, mechanical deformations on the cells are induced by mechanisms such as direct microinjection,[86] cell squeezing,[87] shear stresses,[88] deterministic mechanoporation (DMP),[89] laser optoporation,[90] microfluidic cell hydroporation,[91] vortex shedding,[92] and micro-electric fields.[83] These extensively studied mechanisms temporarily perforate the cell membrane, enabling intracellular delivery. Microfluidics offers several advantages over conventional methods, such as requiring low sample volumes thereby conserving resources, while also maintaining high throughput and transfection efficiency. In cases where there is a limited supply of healthy immune cells from patients, microfluidics can help preserve product quality and facilitate cell manufacturing.

Microfluidics enables the engineering of cells at high-throughput speeds, reaching throughput on the order of million cells per minute.[91,97,98] This capability can address the manufacturing demands of CAR-T yield required for treatments. Microfluidic transfection systems that minimally perturb cell viability are preferred for CAR-T manufacturing as they simplify scale-up production.[99]

Similar to bulk electroporation, microfluidics can be used to deliver various cargoes into cells, including oligonucleotides, complex macromolecules like proteins and carbohydrates, drugs, and dyes.[100] However, a key difference lies in the approach used by microfluidic platforms, which apply membrane-disrupting forces at the microscale, such as rapid mechanical deformations.[101] This microscale approach allows for higher cell viability compared to macro-scale devices like direct microinjection and bulk electroporation. Moreover, as microfluidic techniques are vector free, they are considered safe from any toxicity associated with vector integration and unwanted immunogenicity. This makes microfluidics an attractive option for intracellular delivery in CAR-T cell manufacturing.

Microfluidic electroporation combines microfluidics with electric fields to enhance intracellular delivery, offering a gentler approach compared to bulk electroporation. This technique utilizes thin microchannels, typically measuring tens of micrometers, in either a flow-through or parallel array configuration.[102] Unlike bulk electroporation, which requires voltages in the range of a few hundred volts, microfluidic electroporation enables cargo delivery at much lower voltages, typically in the range of tens of volts. This results in reduced cytotoxicity, making it a promising technique for efficient and less-harmful cell transfection.[103] Initially, the flow-through microfluidic electroporation system was demonstrated by Geng et al., who combined a microfluidic chip with a syringe pump to control the flow rate and electrodes attached to an external power supply to generate electric fields. Their system achieved promising results in plasmid eGFP delivery into CHO-K1 cell lines, with a transfection efficiency of up to 75%, indicating potential for high-throughput applications.[104] Building upon this technology, Choi et al. developed a flow-through microfluidic electroporation system capable of transfecting cells at a rate of 20 million cells per minute, achieving transfection efficiencies of up to 80% in primary T cell-expressing CD19-specific CAR mRNA.[97] Lissandrello and research team developed a microfluidic continuous-flow electrotransfection device that was able to transfect up to 500 million cells per run at a transfection rate of 20 million cells/min. Their microfluidic continuous-flow system reduces the electric field strengths needed for the transfection, thereby reducing the associated Joule heating and toxic effects on cell viability. In their research, they report a transfection efficiency of mCherry encoding mRNA into primary T cells up to 74%.[105] In a later study from the same group published recently, their throughput was further improved to 160 million cells/min, and the versatility of the system was elevated to delivery of gene-editing CRISPR/Cas9 RNPs, eGFP plasmid DNA, and mCherry encoding mRNA into primary T cells at a very high cargo concentration.[106] However, it's important to note that the observed high transfection efficiency and throughput may partly result from the use of high-concentration reagents, potentially increasing manufacturing costs. Microfluidic systems offer precise control over cargo injection into cells and show promise for precise, automated, and high-throughput transfection. Apart from primary T cells, Selmeczi et al. reported successful DC transfection using a microscaled hybrid device featuring parallel stainless-steel-mesh electrodes perpendicular to the fluidic flow direction.[107] This device achieved transfection efficiencies of up to 74% in primary T–cell-expressing eGFP mRNA, with a viability of 90% and a processing rate of 4 million cells per minute, using an extremely high cargo concentration of 200 µg/ml.[107]

A relatively recent technique relies on passing cells through constricted microsized-channels of widths lesser than the cells' diameters in order to squeeze them through the channels utilizing pressure-controlled gradient-driven flow (Figure 4.2a). Sharei et al.[98] showed that a vector-free microfluidic cell squeezing platform was able to deliver fluorescent dextrans into several primary murine-derived immune cells such as B cells, T cells, macrophages, and DCs. Cell squeezing maintains cell viability more than 85% even at high speeds of constricted flow through the microchannels.[98] Ding et al.[102] achieved an impressive 100,000–500,000 cells per second transfection yield from a single microfluidic chip for nucleic acid delivery by employing mechanical and electrical disruption of cell membranes. Moreover, Belling and colleagues[94] integrated acoustic sonoporation, a technique utilizing ultrasound waves, with a microfluidic device to enable high-throughput transfection of primary stem cells and Peripheral Blood Mononuclear Cells (PBMCs), achieving an enhanced throughput rate of 200,000 cells per minute (shown in Figure 4.2b) Further increase in the throughput and efficiency of the device was possible with additional parallel microchannels in the chip and repeated squeezing, respectively. Another work on primary immune cells was reported by Li et al., [101] who devised the microfluidic cell squeezing to introduce Janus kinase (JAK) inhibitors into human primary PBMCs and eventually inhibiting up to 90% of the JAK pathway. In another pioneering study, DiTommaso et al. compared the changes in gene expression profiles between microfluidic

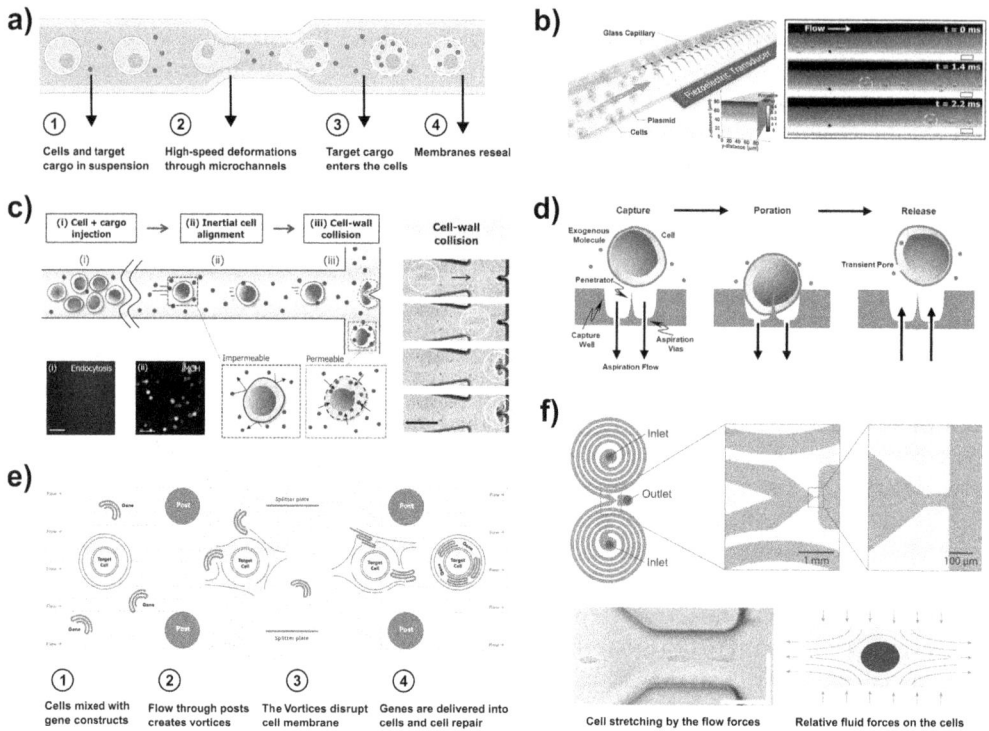

Figure 4.2 Microfluidic transfection system for CAR-immune cell engineering. (a) Schematic overview of microfluidic cell squeezing technology. Cells and cargo pass through microfluidic constrictions of widths lesser than the diameter of cells leading to membrane disruption and subsequent cargo transfection. Adapted with permission.[93] Copyright 2022, Elsevier. (b) Schematic of acoustic sonoporation which uses ultrasound waved parallel to the flow of cells through the microfluidic channels to permeabilize the cell membrane for intracellular delivery. Reproduced with permission.[94] Copyright 2020, The Authors, published by the National Academy of Sciences. (c) Schematic of inertial microfluidic cell hydroporator (iMCH). High-speed microfluidic control of cell-wall collision and fluidic-shear stresses create nanopores allowing cargo to diffuse through the cells for transfection. The fluorescent images show the delivery via endocytosis and iMCH of Fluorescein isothiocyanate (FITC)-dextran into MDA-MB-231 cells (scale bar: 50 μm). Adapted with permission.[91] Copyright 2018, American Chemical Society. (d) Illustrated concept of deterministic mechanoporation demonstrating a single capture site. T cells are first immobilized into the capture wells using a microfluidic negative aspiration flow and then impaled with a single needle-like structure. Intracellular delivery of plasmid DNA occurs via simple diffusion through the transient plasma membrane pore. A reverse fluidic flow can subsequently recover the cells. Reproduced with permission.[89] Copyright 2019, American Chemical Society. (e) Illustration of the microfluidic vortex shedding intracellular delivery mechanism. The technology creates alternate hydrodynamic low-pressure regions or vortices when a fluid is passed through microscale posts. As cells and cargo flow through these vortices, they can disrupt the cell membrane and enable intracellular delivery. Adapted with permission.[95] Copyright 2021, The Author(s), published by Springer Nature. (f) Schematic of device layout of viscoelastic mechanoporation which exploits the cell contraction–stretching behavior due to microfluidic control flow to improve transfection efficiency. Adapted with permission.[96] Copyright 2024, The Author(s), published by Springer Nature.

cell squeezing and non-viral industry standard, bulk electroporation after PD-1 knockout in primary T cells.[108] The authors claim that bulk electroporation upregulated the expression of pro-inflammatory cytokines such as IFN-γ and IL-2 by 600 and 30 folds, respectively, while microfluidic squeezing showed minimal changes in protein expression changes relative to the controls. They also report that CAR-T cells engineered using microfluidic cell squeezing can offer better tumor clearance and antitumor activity in mice compared with bulk electroporation and non-transfected controls, suggesting that microfluidic cell squeezing better preserves CAR-T cell functionalities in vivo.[108]

Numerous other microfluidic designs have surfaced in recent years utilizing the mechanical shear stresses in microfluidic channel for transfection such as optoporation,[90] microfluidic cell hydroporation,[91] and microfluidic vortex shedding.[92] High-intensity laser-induced delivery has also been reported to aid in microfluidic transfection of T cells. Rendall et al.[109] developed a microfluidic platform integrated with a high-intensity laser beam to transfect HL60 cells. This platform attained a modest efficiency of 20% with a membrane-impermeable dye, propidium iodide. Another group reported a similar laser-dependent microfluidic platform capable of transfecting human fibroblasts at a high-throughput transfection rate of a million cells per hour, indicating the throughput capacity of optoporation.[90] However, scale-up of T cell manufacturing might face challenges due to the high wattage required for laser production, and further research is needed to validate its applicability to primary T cells and CAR-T engineering.

Shear stress–induced cell deformation was shown by Meacham et al.,[110] who combined acoustic shear mechanoporation and electrophoresis-induced transfection to introduce large dextran polysaccharides into PBMCs >80% efficiency and plasmid DNA into Jurkat cells with more than 20%. Leveraging shear stresses during microfluidic flow, Hur et al. developed an intracellular delivery device called the microfluidic cell hydroporator (like that shown in Figure 4.2c) This contemporary device utilizes inertial fluidic flow to create precise shear stresses, inducing membrane deformation and enabling cell transfection. Microfluidic cell hydroporation demonstrated high performance, achieving ~90% delivery of GFP mRNA into primary T cells at a remarkable processing rate of 1 million cells per minute per chip.[111] Notably, the Hydroporator™ technology has been commercialized by MxT Biotech for cell engineering applications. It offers high scalability and boasts an extremely low manufacturing cost of about $1 per chip.

Earliest works involving mechanoporation used direct microinjection with a sharp microneedle tip to precisely inject biomolecules into immobilized cells. Then, Adamo and Jensen designed a direct microfluidic injection technique that uses microfluidic flow to control cell interaction and induce mechanical shear for cell transfection. The authors show that their technique can offer transfection at a single-cell resolution up to a few cells every cycle. The implacable drawback lies with the incredibly low throughput of microinjection even though it can ensure a high transfection efficiency.[86] To mitigate the throughput issues with microinjection, Dixit et al. developed a novel microfluidic delivery approach referred as DMP. DMP works by directing primary human T cells into capture wells using microfluidic control where they are impaled using a single nanoneedle-like structure (Figure 4.2d). After the mechanical poration, the plasmid DNA cargo is passively allowed to diffuse into the cells via the transient pore on the cell membrane. The group validated the technique, ensured a high transfection efficiency up to 82% in primary T cells, and improved throughput up to ~7000 cells per run (~71% capture efficiency in 10^4 available capture sites) using massive parallelization of the capture sites and recovery of the mechanoporated cells.[89]

There is limited research on primary T cells conducted using other microfluidic techniques. Microfluidic vortex shedding is another innovative approach utilizing microfluidic control to facilitate intracellular delivery of biomolecule cargo into T cells. This technology relies on creating alternate hydrodynamic low-pressure regions when a fluid is passed through microscale posts (Figure 4.2e). As cells and cargo flow through the fluidic system, these hydrodynamically created vortex regions can transiently perforate cell membranes, enabling intracellular delivery.[92] Jarrell et al. demonstrated that their microfluidic device could deliver eGFP-encoding mRNA into primary T cells with up to 63.6% efficiency.[92] Furthermore, their group in a later study reported successful delivery of gene-editing RNPs targeting the TCRα constant region gene locus with approximately 25% knockout efficiency.[92] Vortex shedding achieves delivery rates of about 2 million cells per minute, with minimal loss of cell viability and minimal changes to early activation and exhaustion marker expression postdelivery. However, this technique requires very high cargo concentrations, up to 160 µg/ml.[92] Sevenler and Toner devised a method termed *viscoelastic mechanoporation* and used high-throughput continuous flow for transfection into primary T cells (Figure 4.2f). Their microfluidic flow device is capable of transfecting dextran-FITC into primary T cells up to 80% efficiency and TCR knockout efficiency of 53% with Cas9 RNPs.[96] These advancements highlight the potential of microfluidic-based techniques for efficient and high-throughput intracellular delivery in T cell engineering and other cell-based therapies. However, further research and optimization are needed to fully exploit the capabilities of these technologies for large-scale manufacturing of genetically modified T cells.

Despite the promising efficiency and throughput of microfluidic transfection, the transfection efficiency and applicability in primary immune cells have been pretty low. Despite the high efficiency and throughput of microfluidic transfection, the operational expenses for cell manufacturing can rapidly increase. The cargo suspended in the microfluidic channels must often be highly concentrated in the range of 20–200 µg/ml. Such high concentrations are necessary to ensure an adequate quantity of cargo enters the cells, resulting in very high transgene expression levels. During transfection, the cargo is free floating in the microfluidic chambers that can lead to cargo degradation and considerable amounts of cargo wastage, adding onto the manufacturing costs. Another drawback of microfluidic cell deformation is the variation in transfection observed among heterogeneous cell populations. Microfluidic channels are designed with uniform widths, causing smaller cells to deform less compared to larger cells. This variation significantly impacts transfection efficiency and cell viability, further reducing final product quality. A critical challenge with microfluidic transfection is microchannel clogging of the microfluidic channels with cells, dust, or cargo. Clogging generally arises from boundary restrictions during the fluidic flow and can significantly affect cell quality, throughput, and efficiency of the technique. This challenge becomes even more pronounced when dealing with low cell sample volumes from patients, limiting the overall manufacturing throughput. For instance, Sharei et al. reported that their microfluidic cell squeezing platform faced channel clogging for every million cells processed and needed de-clogging every 50 seconds. SQZ Biotechnologies Company founded a decade back was at the forefront of microfluidic technologies for CAR-T engineering in clinical trials. Cofounded and led by Langer and team at Massachusetts Institute of Technology (MIT), their proprietary Cell Squeeze technique was reportedly entering phase I/II trials for CAR-T engineering against human papillomavirus (HPV)16+ solid tumors with support from a big pharmaceutical company. But ridden with clogging issues and finding solutions to improve the throughput, the company in March 2024 announced dissolution and sale of all its assets. Belling et al.[112] demonstrated the use of lipid bicelles to coat the microchannel walls to minimize cell attachment and clogging. But the additional use of de-clogging material could quickly increase production cost.

Transfection with nanoparticles

In the past few decades, nanoparticles have risen in popularity for their applications in different sectors such as drug delivery,[113] biomedical diagnostics,[114] antimicrobials,[115] and biotechnology.[116] Nanoparticles are materials with sizes falling in the range of nanometers to a few micrometers. They are easy to synthesize and are amenable to modifications to their size, shape, and compositions. Due to these advantages, nanoparticles have also been used extensively for other applications in adoptive therapy to improve manufacturing, T cell activation, and functionality during T cell therapy. We refer the readers to excellent reviews that focus on these explored approaches.[117–119] In this chapter, we will focus only on the use of nanoparticles for CAR-T transfection and engineering.

In recent years, they have been applied for intracellular delivery of bimolecular cargo like nucleic acids, peptides, antibodies, and drugs. Nanoparticles have shown promise for in vitro delivery to primary cells and cell lines as well as in vivo applications to tissues and organs.[120–122] Cells uptake nanoparticles using several endocytic mechanisms like pinocytosis, phagocytosis, and clathrin-mediated and caveolae-mediated pathways. In NK cells, it was reported that particles of different sizes are endocytosed using different pathways.[123] In general, particle sizes <100 nm diameter induce clathrin-mediated pathways, while particles of larger sizes utilize caveolae-mediated pathways and pinocytosis.[123] Among different immune cell types, only those involved in active phagocytic activity like neutrophils and macrophages endocytose extracellular materials. Therefore, often nanoparticle entry into other relevant immune cells like T cells needs to be synthesized from materials that promote surface adhesion and endocytosis. For intracellular delivery, nanoparticles typically encapsulate cargo biomolecules, acting as cargo-loaded nanocarriers to the immune cells. Once nanoparticles are endocytosed by immune cells, they become trapped within acidic endosomes. However, they are released from these endosomes before reaching the intended target sites within the cells.[124,125]

Similar to other non-viral techniques, nanoparticles can be used for intracellular delivery of almost any kind of biomolecule cargo as long as they are able to be attached to the nanoparticles through electrostatic interactions or lipid–bilayer fusion, or physical entrapment. In literature, nanoparticles are frequently used with different formulations to deliver DNA into various immune cells such as T cells, NK cells, DCs, and macrophages for immunotherapy. For instance, catatonically charged lipid nanoparticles (LNPs) and liposomes are used to encapsulate DNA within their lipid cores. Due to the negative charge of nucleic acids, the electrostatic attraction helps in efficient and stable encapsulation and provides protection from degradation within acidic endosomes. Small oligonucleotides like siRNA were reportedly encapsulated within lipid-based envelope-type nanoparticles to induce gene editing of more than 75% in different immune cell lines such as Jurkat T cells, THP-1 monocytes, KG-1 macrophages, and NK-92 cells.[129] Manganese dioxide NPs delivered TGF-β receptor 2 targeting siRNA into ~92% NK-92 cells and successfully induced about 90% knockdown of the mRNA activity.[130]

Billingsley et al. demonstrated that ionizable LNPs can effectively deliver CD19 CAR mRNA to primary human T cells with significantly lower toxicity compared to bulk electroporation[126] (Figure 4.3a), although the CAR expression efficiency and in vitro tumor killing efficacy of liposome-transfected primary T cells were comparable with bulk electroporation. This research work exemplifies the potential of LNPs in mRNA-based CAR-T therapy. But there was notable cytotoxicity observed in KG-1 and NK-92 cells treated with LNPs, and Xue et al. explored that this was a result of cell membrane rupture upon interaction with the cationic lipid head. This research work warrants further enhancements to mitigate nanotoxicity.[131] Non-toxic nanoparticles made from polymers are considered as popular carriers to mitigate nanotoxicity in immune cells. Yu et al. used self-assembled nanoparticles for delivering EGFRvIII CAR plasmids for in vitro CAR-T engineering and reported a high CAR expression in Jurkat T cells. Furthermore, the engineered CAR-T cells exhibit lower cytotoxicity and specific tumor lysis[127] (Figure 4.3b). The synthesis of self-assembling nanoparticles is quite simple, and they are known to offer controlled release.

Smith and colleagues achieved a significant milestone by generating CAR-T cells targeting CD19+ B cells in vivo using biodegradable, cationic, poly(ß-amino ester) (PBAE)–based nanoparticles for the first time (similar to the work shown in Figure 4.3c). To enhance delivery of the plasmid into the nucleus, which most DNA transfection techniques need for efficient transgene expression, they functionalized PBAE with peptides composed of microtubule-associated sequences and nuclear localization signals, which facilitate the microtubule transport mechanism to enter the cell nucleus. After injection of the nanoparticles in a mouse leukemia model, they observed robust T cell uptake and prolonged CAR transgene expression for at least two weeks, while the average survival rate was improved by 58 days in the treated mice.[132] Nevertheless, the programmability of the nanoparticles was not flawless. Smith et al. additionally observed that, apart from T cells, they observed off-target effects, as nanoparticles were found to be internalized by circulating neutrophils and monocytes, although their numbers were significantly in lower quantities compared to the target T cells.

Beyond cancer immunotherapy, lipid-based-nanoparticle-modified CAR-T cells were used in vivo to treat cardiac injury preventing fibrosis progression and rescuing heart function. Rurik et al. showed that up to 24.7% CAR-T cell population was observed in the treatment mouse group injected with nanoparticles encapsulating fibroblast activation protein–specific CAR mRNA.[133] Nanoparticle-based CAR-NK cells were generated by Kim et al., who reported a 60% transfection efficiency in vitro in NK-92MI cell line.[134] Their Polyethyleneimine (PEI)-coated magnetic nanoparticles induced significant antitumor activity in vitro and caused no NK cell toxicity.[134] Wilk et al.[135] interestingly designed a charge-altering releasable transporter (CART) system for targeting primary human NK cells. The CART system was capable of transfecting more than 10% of the NK cell population, significantly higher than the commercially used lipofectamine. Engineered CD19 targeting NK cells also expressed cytotoxic markers such as CD107a, TNF-α, and IFN-γ, and exhibited a high cytotoxic activity against tumor cells in vitro. CAR mRNA delivery to murine primary macrophages and tumor cell killing was recently reported by Ye and group using LNPs. There was a slight transfection difference between primary Bone marrow-derived macrophages (BMDM) transfection (51.1%) and M1 macrophages (60.7%), which resulted in B-cell lymphoma eradication in vitro by 32.54% and 22.50%, respectively.[136] Cruz and colleagues demonstrated that poly(lactic-co-glycolic acid)-coated nanoparticles facilitated the delivery of antigens into DCs, which subsequently stimulated CD8+ T cells.[137]

Over the couple of years, NP-based CAR therapy strategies have demonstrated significant growth and promise. One notable advantage of nanoparticles is their ability to remove the immunogenicity concerns associated with viral vectors and biological cell

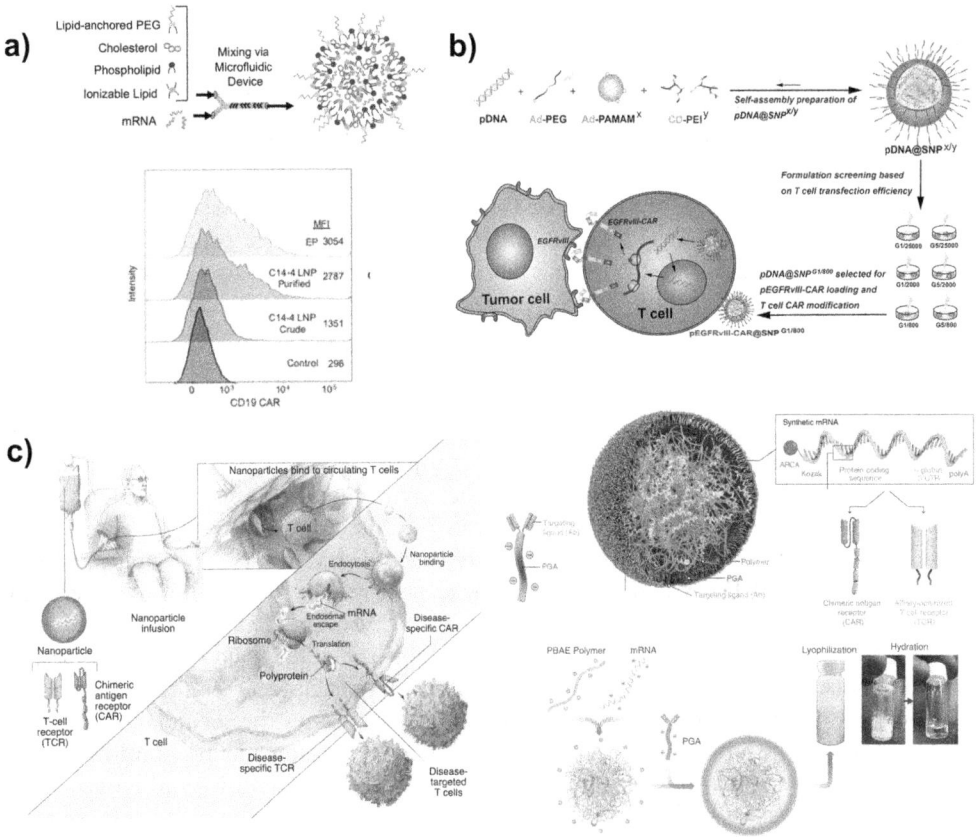

Figure 4.3 Nanoparticle-based transfection system for CAR-immune cell engineering. (a) Ionizable lipid nanoparticles (LNPs) deliver mRNA into T cells and induce successful CAR expression. Size (z-average) distribution revealed an average diameter of 70 nm. The flow histogram microimages demonstrate that the purified LNPs loaded with CAR mRNA induced a similar mean fluorescence index of CAR expression to the industry standard, bulk electroporation (EP), and significantly higher than control and crude LNPs. Adapted with permission.[126] Copyright 2020, American Chemical Society. (b) Self-assembled nanoparticles loaded with plasmid DNA show CAR gene expression on T cells. Schematic representation showing the preparation and formulation screening process of plasmid DNA-loaded self-assembled nanoparticles (pDNA@SNPs) for expression of pEGFRvIII-CAR and CAR-T cell cytotoxicity. Reproduced with permission.[127] Copyright 2020, Dove Press. (c) Schematic showing the steps involved in the in situ administration of mRNA-encapsulated polymeric nanoparticles to engineer T cells to express specific CARs. The cargo mRNA was loaded onto nanoparticles prepared using biodegradable poly(ß-amino ester) (PBAE) as the carrier matrix and coated with polyglutamic acid and anti-CD8 antibody ligand. These particles are coated with anti-CD8 ligands and ensure a target delivery to cytotoxic T cells, and once they are infused into the patient's circulation, they can transfer the transgenes they carry into the lymphocytes and transiently reprogram the T cells. Reproduced with permission.[128] Copyright 2020, The Author(s), published by Springer Nature.

perturbations associated with bulk electroporation.[138] Despite considerable interest, NP-based CAR strategies still lag behind viral vectors in terms of understanding the intracellular endosomal release machinery. Even though more than 80 LNP-based products are in clinical development, nanoparticles are poorly controllable in terms of cargo release.[139] Notably, with Onpattro® (patisiran), the U.S. FDA–approved LNP-based siRNA drug, only 1%–4% of nucleic acids encapsulated within the LNPs escape the acidic endosomes and reach cytoplasm.[140,141] Several other studies also indicate that less than 2% of the internalized cargo reaches the cytoplasm.[142,143] In addition to endosomal degradation, exocytic release mechanisms from the cells, or degradation by cytoplasmic endonucleases, and autophagic clearance serve in reducing the gene delivery efficiency even further.[142,144]

Hepatic aggregation poses a significant extracellular challenge for nanoparticle-based therapies, with lipid-based nanoparticles accumulating in the liver at high rates, ranging from 30% to 90%. This accumulation increases the risk of off-target effects, particularly in the treatment of non-hepatic diseases.[145] Nanoparticles are typically internalized by cells within acidic endosomes, where the acidic environment can degrade nucleic acids such as DNA, mRNA, and siRNA. While strategies like establishing pH-sensitive links between nanoparticles and cargo can minimize cargo degradation, regulating the intracellular transport of DNA cargo into the nuclei post-release from endosomes remains challenging and is crucial for transgene expression.[146] Another challenge associated with nanoparticle utilization is cytotoxicity, especially with metallic solid nanoparticles and cationic lipids. In vivo administration of metallic nanoparticles can lead to in vivo accumulation and organ toxicity.[147] A recent review also discusses the secondary impact of nanoparticles on the human body and the impact on in vivo nanomedicine.[148] Additionally, the use of LNPs with cationic lipid heads can damage cell membranes and induce cytotoxicity and cell lysis at high concentrations.[131] To address issues of inadequately controllable intracellular cargo release and transport, nanoparticles are often added at elevated concentrations to immune cells, resulting in alterations in gene expression. Studies indicate metallic, metal oxide, and carbon nanoparticles induce oxidative stress, apoptosis, necrosis, and metabolic imbalances.[149-151] To mitigate immunogenicity and improve biocompatibility, surface functionalization of nanoparticles has been explored. For instance, Mostaghaci et al. functionalized the surface of calcium phosphate nanoparticles with amine groups, enhancing interactions with cells and resulting in concentration-dependent transfection efficiency, immune activation, and cytotoxic effects.[152] These approaches aim to improve the safety and efficacy of nanoparticle-based therapies and mitigate adverse effects associated with their use.

There is a critical need to investigate how nanoparticles influence the non-specific proliferation or inhibition of targeted immune cells and their impact on therapeutic interventions. Recent research has demonstrated that nanoparticle-mediated transfection significantly affects immune cell proliferation and is heavily dependent on the type of nanoparticles used.[153] Furthermore, it has been observed that immune cells from aged patients exhibit reduced uptake and transportation of nanoparticles, while also experiencing greater nanoparticle-induced biological disruptions. However, transfection of immune cells from older patients, who constitute the majority of cancer patients, remains inadequately studied. As a result, achieving successful primary immune cell transfection using nanoparticles has yielded only moderate success rates of ~10%–30%. This intracellular barrier impedes the efficient delivery of nucleic acids into cells and serves as a critical rate-limiting factor for in vivo–targeted delivery and successful intracellular expression.

Further research into CAR-T applications and an in-depth comprehension of nanoparticle–cell interactions with cells will facilitate the development of effective CAR-based anticancer medications. Nevertheless, with the recent clinical approvals for gene therapy utilizing LNPs along with the recent success of LNP-based SARS-CoV-2 vaccines, the field of nanoparticles is anticipated to rapidly evolve and emerge as a promising non-viral transfection technology.

Transfection with high-aspect-ratio nanostructures

High-aspect-ratio nanostructures are one-dimensional, well-defined structures with an aspect ratio (height:width) greater than ten, which can physically penetrate cells through interactions with the cell membrane to achieve transfection. Due to the recent advances in nanofabrication technologies such as nanolithography, thin film deposition, and dry/chemical etching, a range of nanostructures have been designed and developed for intracellular delivery of biomolecule cargo. In contrast to nanoparticles, which primarily rely on endocytosis, high-aspect-ratio nanostructures induce transient membrane perforation to enhance intracellular cargo release. Nanostructure-mediated transfection operates as a vector-free system, eliminating the need for additional carrier encapsulation steps and thereby minimizing production costs on a large scale. As a non-viral transfection tool, nanostructures also mitigate virus-mediated safety risks and immunogenicity. Moreover, unlike bulk electroporation, nanostructures obviate the necessity for extremely high voltages (~500 V), thereby preserving biological cell functions post-transfection. Similar to other physical transfection techniques, high-aspect-ratio nanostructures are capable of transfecting various cargo types, including oligonucleotides, peptides, and larger nucleic acids. This versatility makes them invaluable for transfection scenarios requiring simultaneous delivery of multiple cargo types at high efficiency. By leveraging localized interactions at the nanostructure–cell interface, precise control of cargo delivery can be achieved. Furthermore, due to the localized nature of interactions at the cell interface (ranging from 10 to 400 nm), transient perforation induced by nanostructures minimally invades cells and importantly preserves cell viability.

Xie et al.[154] reported that transient pores formed on cells during nanostructure-mediated transfection reseal within 10 minutes, as demonstrated by tracking the entry of impermeable dye into cells immediately after transfection. Nanostructures have been successfully implemented in transfection of challenging cell types such as human-induced pluripotent stem cell–derived cardiomyocytes,[155] primary human immune cells,[156] and primary human neuronal cells.[157,158] Diverse high-aspect-ratio nanostructures, characterized by diverse sizes, shapes, and materials coupled with electrical and magnetic properties, have been engineered for various applications, including electrophysiological recordings, biochemical sensing, stimulation, and intracellular extraction. For detailed insights into these areas, we recommend referring to comprehensive reviews on nanofabrication and several applications of nanostructures.[68,159] In this chapter, we focus only on the use of high-aspect-ratio nanostructures for immune cell transfection.

Researchers can ensure efficient cargo loading, cell interface interactions, and successful cargo release intracellular delivery by modifying the physical parameters of vertical nanoneedles and nanowires. The adaptable topological configurations of nanoneedles facilitate versatility in loading, co-loading, and releasing various types of cargo.[164–166] Optimization of cell penetration can be achieved through manipulation of high-aspect-ratio nanostructure geometry (density, length, and diameter), surface functionalization, and duration of nanoneedle-cell interfacing.

Generally, vertically aligned nanostructures, referred to as nanowires[167] or nanoneedles,[168] have a sharp tip capable of impaling the cell membrane spontaneously. Cargoes are coated on the tip of solid nanoneedle structures, where cargo adsorption onto the solid nanoneedles is facilitated by electrostatic interactions resulting from surface functionalization. In some cases, cargoes can be added to the cell culture medium to increase cargo availability during intercellular delivery. Cell interactions with nanowires can occur through passive settling and adhesion, or through the application of an external force such as centrifugation. However, the exact mechanism by which nanostructures mediate penetration and intracellular delivery remains a subject of ongoing debate.[169,170] Shalek et al. coated biomolecule cargoes such as plasmids, siRNA and peptides onto surface of vertically aligned silicon nanowires subsequently utilized the nanowires to achieve transfection of different primary immune cells, including DCs, B cells, T cells, macrophages, and NK cells[160,167] (Figure 4.4a). In resting murine CD4+ T cells, they report a 77% gene knockdown efficiency using nanowire-mediated injection. In a later study, the same group report nanowire-mediated siRNA delivery affected the dynamic regulatory network of Th17 differentiation, notably without compromising the phenotype of unstimulated T cells. It was previously proposed by Xie et al. that in the presence of centrifugal forces, the cell membrane continuously warps around the nanowires inducing localized membrane tension and large-scale deformations. These deformations eventually cause membrane rupture or perforation. But many other mechanisms for penetration have also been discussed before.[159,171] The lack of a cargo reservoir and limitation of the amount of cargo that can be coated over the nanoneedle surface makes solid nanoneedles suitable for a single use, one-time delivery system.[172]

Recently, porous nanoneedles have been developed, providing precise cargo absorption within the nanoneedles and subsequent release upon interfacing with cells.[168] Porous nanoneedles can be customized with a larger surface area, enabling them to load higher amounts of cargo within their structures. It has been reported that the large amounts of cargo encompassed within the needles can facilitate prolonged delivery of concentrated cargoes into target sites for hours to weeks, rather than a rapid, one-time injection.[173] Cargo can be loaded into porous nanoneedles through physisorption, with porous needles amassing over 300 times the payload compared to solid needles.[174] Additionally, porous needles have demonstrated biocompatibility and biodegradability during cell interfacing. In contrast, mesoporous silicon-based nanoneedles are better suited for in vivo applications due to their bioresorbable nature and ability to provide controlled release of transfection cargoes.[175,176] However, there is limited research on the applicability and efficacy of nanoneedles with primary immune cells. Further investigation is necessary to understand the biology of immune cell transfection utilizing both solid and porous nanoneedles.

In the past decade of research, direct penetration by nanostructures has been observed to be a rare event. Consequently, coupling nanostructures with external forces such as centrifugation[162] or nanoelectroporation[154] has become the most commonly reported method for transfecting immune cells to enhance delivery efficiency. This has led to innovations in nanostructures such as nanotubes, [177] nanocones,[178] or nanostraws[154] assisted with external mechanical, centrifugal, electrical, or magnetic forces to enhance cell membrane permeabilization.

Alternatively, Hollow nanoneedles or nanotubes,[177] which are "hollow" versions of nanoneedles, are utilized to load cargo amounts in a controlled fashion within the inner

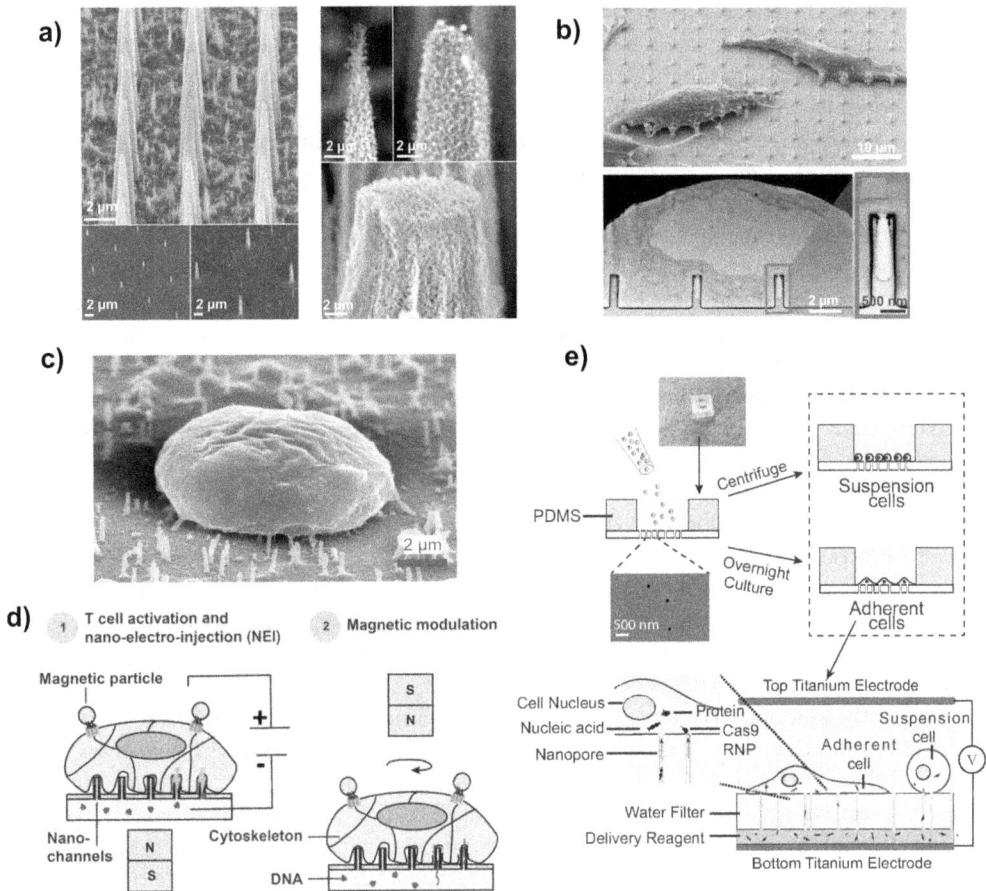

Figure 4.4 High-aspect-ratio nanostructures for CAR-immune cell engineering. (a) Scanning electron microscopy (SEM) images showing the use of solid nanowire mechanoporation of primary immune cells. Biomolecule transfection is observed in murine bone marrow–derived dendritic cells (BMDCs), B cells, dendritic Cell (DCs), macrophages, natural killer (NK) cells, and T cells. Reproduced with permission.[160] Copyright 2012, American Chemical Society. (b) SEM image showing the top view of cells interfacing gold (Au)-coated nanotubes for electro-active nanoinjection (ENI) of biomolecule cargo into the cells. Cross-sectional view of the cell ENI platform-cell membrane interface is shown in the SEM image below at different magnifications. Reproduced with permission.[161] Copyright 2023, The Author(s), published by Springer Nature. (c) SEM image of the nanostructure-interface showing primary hematopoietic stem progenitor cells (HSPCs) on top of nanostraws. Centrifugation enhances the close adherence of HSPCs to nanostraws and promotes transfection with nanoelectric pulses. Reproduced with permission.[162] Copyright 2020, The Authors, published by National Academy of Sciences. (d) Schematic representation of nano-electric injection using magnetic field strengths. Magnetic fields are used to direct magnetic-bead bound cells toward the nanostraws. And then, nanoelectric pulses are used to electrophoretically conduct biomolecule transfection into T cells. Reproduced with permission.[156] Copyright 2021, Wiley-VCH GmbH. (e) Schematic of nanopore-electroporation technique utilizing track-etched nanoporous membranes for biomolecule transfection into both suspension and adherent cells. Reproduced with permission.[163] Copyright 2019, The Authors, published by National Academy of Sciences.

cavity without any need for surface functionalization. Nanotubes offer a larger loading surface area compared to solid nanoneedles, and with externally applied forces like centrifugation, electrical, magnetic, or user-applied forces, they can be used for transfection. These devices have large nanotube silicon arrays with specific mechanical properties that are instrumental in achieving greater control over cargo loading and release into the cells through diffusion. Chen et al. designed arrays of vertically aligned silicon nanotubes that can be incorporated with a centrifuge to deliver CAR constructs into primary mouse T cells.[177] Suspending primary mouse T cells under a centrifugal force of 200 × g for 15 minutes was sufficient to transfect ~37% CD19 CAR constructs into the cells, and subsequently demonstrated enhanced proliferation and cytotoxicity against CD19-expressing tumor cells.[177] To necessitate further improvement to the relatively low efficiency, in a late study the same group implemented low voltage electrical fields into the nanotube arrays. The work led by Elnathan reported a two-fold increase to 68.7% transfection efficiency of CD19 CAR constructs in primary human T cells the with 10V electric pulse train (Figure 4.4b). The resulting CAR-T cells exhibited significantly improved cytotoxic functions and cellular functionalities compared to bulk electroporated CAR-T cells.[179] Shokouhi et al. also demonstrated their electroactive nanoinjection platform activity in GPE-86 mouse fibroblasts where they achieve ~52% plasmid delivery, ~55% mRNA delivery, and about 40% knockdown efficiency induced by siRNA delivery.[161] Their group's electroactive hollow nanotubes enable precise dosage control for cargo release and delivery in response to electric-field modulation, whereas cargo release in hollow nanotubes relies on diffusive processes.

A recently popularized nano transfection technique called nano-electric injection (NEI) combines the use of hollow nanostructures called nanostraws with highly localized electric fields generated to create transient membrane pores on the cell membrane and directly inject charged cargo species into targeted cells.[154] This platform enables precise dosage control by adjusting reagent concentrations, nanostraw geometries, electrical pulse duration, and voltages parameters.[155] The nanostraws are usually developed on thin polymer material (polycarbonate or polyester) connecting a microfluidic chamber serving as a cargo reservoir to the cells, and thus can mitigate the challenges of limited cargo availability during the transfection. Unlike bulk or microfluidic electroporation, the highly localized electric fields can reach 2–8kV/cm at the cell–nanostraw interface (nanochannels ranging 100–200 nm) from just the 20–40 V applied within the circuit. Nanostraw electroporation uses voltage strengths of small magnitude, thus minimizing the biological perturbations, and offers higher cell viabilities >90%, and transfection rates >80% for several cargo types.[68,154–156] Tay and Melosh reported that cells transfected with NEI had minimal changes to the cell proliferation rates, cell stresses, and global gene expression, while bulk electroporated cells suffered from 40% delay in cell proliferation times and significant changes to global gene expression profiles.[68] He et al. also implemented coating of anti-epithelial cell adhesion molecule (EpCAM) onto nanostraws to capture circulating tumor cells and subsequently utilized the nanostraw electroporation platform to deliver biomolecules into the cells over a period of time.[180] Another study by Choi et al. demonstrated that a carbon nano-syringe array can achieve higher transfection efficiency up to ~46% EGFP plasmid when assisted with centrifugal force. But, Choi's described method uses very high centrifugal forces (1600 × g) that could negatively impact the cell viabilities after transfection.[181] Another promising delivery method uses nanostraw stamping, which uses an array of nanostraw stamps loaded with delivery cargo. The stamp is pressed against the cells to physically permeate them and release cargo via the nanostraws. For instance, Zhang et al. introduced a method for directly delivering molecules into adherent cells using an Au-based nanostraw membrane stamp that was capable of injecting oligonucleotides up to 83% efficiency in NIH-3T3 cell line; the method is still yet to be shown to work on primary immune cells.[182] Human hematopoietic stem and progenitor cells (HSPCs) are suspension cells known to be harder to transfect. Schmiderer et al. demonstrated that with a relatively low voltage and pulsed square waveforms, various cargoes such as Dextrans, oligonucleotides, and siRNA were delivered >70% efficiency within just a two-minute electroporation duration.[162] (Scanning electron microscopy [SEM] image of the HSPCs on top of nanostraws is depicted in Figure 4.4c.) Configured with a cargo reservoir beneath the alumina nanostraws, Xie et al. showcased their transfection system can be utilized for co-transfection of biomolecules.[154] Cao et al. designed a nanostraw electroporation system that can achieve a quantitatively controlled co-transfection of plasmids and mRNA into adherent and suspension cells.[155] These research works demonstrate the versatility of cargo delivery using nanostraw platform.

Beyond the use of nanoelectroporation, magnetic fields have also been explored to enhance transfection with nanostructures. One approach reported by Tay and Melosh used magnetic beads to bind primary T cells and later used a strong magnetic field to pull them down toward the nanostraws[156] (Figure 4.4d). The magnetic field strength results in close adherence of primary T cells toward the nanostraws following which a pulsed train of electric fields is allowed to pass through the cells resulting in effective transfection. The efficacy of GFP plasmid delivery was improved to ~50% when combined with a magneto-mechanical stimulation after nanostraw transfection.[156] This magnetic nanoelectroporation method can overcome challenges in production costs and manufacturing delays by removing the need for bulky centrifuges and minimizing human labor. Another example is the nanotube spearing technique by Cai et al., which utilized nickel-embedded carbon nanotubes that can be guided toward cells to puncture them with a magnetic field.[183] Cai et al. report > 85% delivery efficiency of EGFP plasmids into Bal17 cells and primary murine B cells.[183] Xie's recent publication highlights that the majority of physico-chemical techniques combined with nanostraws yield low plasmid transfection efficiency in HL-1 (cardiac muscle) and DC2.4 (mouse dendritic) cells. The need for an integrated microfluidic reservoir to ensure transfection can lead to increased fabrication costs and operational complexity.[184]

High-aspect-ratio nanostructure arrays entail high manufacturing costs and require the use of expensive equipment in fabrication methods such as photolithography, atomic layer deposition (ALD), and plasma reactive ion etching (RIE).[185,186] They are also labor-intensive to prepare and need highly trained personnel to handle the equipment and reduce variation during fabrication. Lard et al. proposed a dielectrophoresis method that directs T cells to the nanostraws, thereby showing a way to reduce labor-intensive methods to pull the T cells toward nanostraws before transfection.[187] Wen et al. came up with an interesting approach to lower cost and simplicity of the nanostraw-electroporation technique. By electroplating the surface of nanostraws to manufacturing conductive nanostraw arrays, their method uses voltages as low as 5V to achieve transfection while maintaining cell transfection capability.[186] Cao and their team addressed the challenges of huge cost by utilizing polycarbonate membranes with nanopores to replace nanostraws during nanoelectroporation[163] (Figure 4.4e). Application of electric voltages ranging between 20 and 60V was used to achieve transfection of mRNA (up to 75%) and plasmids (up to 52%) into suspension Jurkat T cells.[163] A research work led by Espinosa and their group also has employed the nanopore electroporation technology to develop a localized electroporation device (LEPD), which could offer a 24-multi-well transfection system and transfect up to 1.2 million cells in a single run (50K cells per LEPD well).[188] The nanopore electroporation technology yet to be investigated on primary T cells could potentially reduce the complexity behind nanostraw fabrication and offer a cost-effective solution to CAR-T manufacturing. But in a recent article, Xie and their group also signified that the nanostraw transfection of large plasmid DNA into suspension (Jurkat-T) cells is much harder to accomplish with a meager efficiency of 19%, which warrants more research on the applications for large-scale T cell manufacturing with nanopore-electroporation technology.[189]

Nanostructures have achieved high transfection efficiency and cell viability while accessing the cytosol, opening up new frontiers of applications in immune cell transfection and CAR-T engineering. But penetrating the cytosol of a living cell could also pose potential harm such as cytosol and intracellular content leakage from within the cells. As CAR-T therapy gains more traction, tools offering high throughput, efficiency, and minimal cell perturbations would further drive the early technologies into clinical investigations. High-aspect-ratio nanostructures are an appealing alternative for the gold standard T cell transfection tools, although the functional ability of primary immune cells and scalability need to be validated in clinical settings to validate their potential for T cell therapeutics. While the downsides of viral tools and bulk electroporation can be significantly removed using nanostructures, it also becomes imperative to understand how nanofabrication technology can be exploited for precise dosage control into primary immune cells and expand the applicability for large-scale CAR-T engineering.

Discussion and outlook

Over the past decade, CAR-T cell therapy has made significant progress in the clinics, but several limitations still remain unaddressed, thus narrowing its application to specific blood cancers. To overcome the toxic side effects and improve accessibility to a wider range of patients, several engineering solutions need to be concurrently developed to not only target the tumor antigen but also maintain T cell functionalities to keep surveillance of potential tumor reoccurrence. Innovative designs will likely require the simultaneous introduction of transgenes and knockout of multiple genes to achieve multi-antigen specificity, which could be critically relevant in CAR-T therapeutics for treating solid tumors. Recently published research work also signifies that several patient-derived T cell receptors and human leukocyte antigens may need to be silenced to minimize the risk of GvHD. Checkpoint regulators of immune exhaustion and T cell activation must also be disrupted to overcome the immunosuppressive effects of the tumor microenvironment. Development of suitable transfection technologies capable of accommodating complex CAR constructs additionally burdened with genome-editing components and immune effector molecules is critical in achieving this target.

While viral vectors and bulk electroporation are optimally situated in the clinical T cell therapeutics setting due to their high efficiency, we learn that these techniques possess several downsides such as safety and immunogenicity concerns, limited cargo capacity, cell toxicity, and reduced T cell functionalities. These techniques also require specialized GMP facilities, regulatory approvals, and need a huge manufacturing cost and time to produce clinical-grade T cells. To address this manufacturing challenge, significant research efforts have been dedicated to non-viral transfection technologies compatible with the production of next-generation T cell therapies. Besides, in the past few decades, non-viral physical and chemical-mediated transfection approaches are actively being researched for their interactions with homeostasis and long-term functionality of the T cells. We reviewed and presented three major emerging technologies, using microfluidic devices, nanoparticles, and high-aspect-ratio nanostructure arrays, which show the most promise in terms of rapid preclinical success with high potential for entering clinics in the future. To meet the requirements of cell manufacturing, next-generation T cell therapeutics must generate more effective and safer cell products, and thus, while still in preclinical exploration, novel non-viral transfection technologies showing potential for automation can easily be adopted in clinical use.[190] Although there is limited clinical research conducted currently with these emerging tools, there is significant interest observed via startups and spin-offs entering the CAR-T commercial landscape.[191]

References

1. Sung, H. et al. Global cancer statistics 2020: GLOBOCAN estimates of incidence and mortality worldwide for 36 cancers in 185 countries. *CA Cancer J Clin* **71**, 209–249 (2021).
2. Stevanović, S. et al. Complete regression of metastatic cervical cancer after treatment with human papillomavirus–targeted tumor-infiltrating T cells. *J Clin Oncol* **33**, 1543–1550 (2015).
3. Besser, M. J. et al. Clinical responses in a phase II study using adoptive transfer of short-term cultured tumor infiltration lymphocytes in metastatic melanoma patients. *Clin Cancer Res* **16**, 2646–2655 (2010).
4. Rosenberg, S. A. et al. Durable complete responses in heavily pretreated patients with metastatic melanoma using T-cell transfer immunotherapy. *Clin Cancer Res* **17**, 4550–4557 (2011).
5. Rosenberg, S. A. & Restifo, N. P. Adoptive cell transfer as personalized immunotherapy for human cancer. *Science (1979)* **348**, 62–68 (2015).
6. Voss, R.-H. et al. Molecular design of the Cαβ interface favors specific pairing of introduced TCRαβ in human T cells. *J Immunol* **180**, 391–401 (2008).
7. Tsimberidou, A.-M. et al. T-cell receptor-based therapy: an innovative therapeutic approach for solid tumors. *J Hematol Oncol* **14**, 102 (2021).
8. Spindler, M. J. et al. Massively parallel interrogation and mining of natively paired human TCRαβ repertoires. *Nat Biotechnol* **38**, 609–619 (2020).
9. Chandran, S. S. & Klebanoff, C. A. T cell receptor-based cancer immunotherapy: Emerging efficacy and pathways of resistance. *Immunol Rev* **290**, 127–147 (2019).
10. van der Stegen, S. J. C., Hamieh, M. & Sadelain, M. The pharmacology of second-generation chimeric antigen receptors. *Nat Rev Drug Discov* **14**, 499–509 (2015).
11. Maher, J., Brentjens, R. J., Gunset, G., Rivière, I. & Sadelain, M. Human T-lymphocyte cytotoxicity and proliferation directed by a single chimeric TCRζ /CD28 receptor. *Nat Biotechnol* **20**, 70–75 (2002).
12. Chmielewski, M. & Abken, H. TRUCKs: the fourth generation of CARs. *Expert Opin Biol Ther* **15**, 1145–1154 (2015).
13. Kochenderfer, J. N. et al. Eradication of B-lineage cells and regression of lymphoma in a patient treated with autologous T cells genetically engineered to recognize CD19. *Blood* **116**, 4099–4102 (2010).
14. Maude, S. L. et al. Tisagenlecleucel in children and young adults with B-Cell lymphoblastic leukemia. *N Engl J Med* **378**, 439–448 (2018).
15. Wang, M. et al. KTE-X19 CAR T-cell therapy in relapsed or refractory mantle-cell lymphoma. *N Engl J Med* **382**, 1331–1342 (2020).
16. Neelapu, S. S. et al. Axicabtagene ciloleucel CAR T-Cell therapy in refractory large B-cell lymphoma. *N Engl J Med* **377**, 2531–2544 (2017).
17. U.S. Food and Drug Administration. Approved Cellular and Gene Therapy Products. (2024).

18. Ruggeri, L. et al. Effectiveness of donor natural killer cell alloreactivity in mismatched hematopoietic transplants. *Science (1979)* **295**, 2097–2100 (2002).
19. Liu, E. et al. Cord blood NK cells engineered to express IL-15 and a CD19-targeted CAR show long-term persistence and potent antitumor activity. *Leukemia* **32**, 520–531 (2018).
20. Rezvani, K. Adoptive cell therapy using engineered natural killer cells. *Bone Marrow Transplant* **54**, 785–788 (2019).
21. Liu, E. et al. Use of CAR-transduced natural killer cells in CD19-positive lymphoid tumors. *N Engl J Med* **382**, 545–553 (2020).
22. Xie, G. et al. CAR-NK cells: A promising cellular immunotherapy for cancer. *EBioMedicine* **59**, 102975 (2020).
23. Kalyan, S. & Kabelitz, D. Defining the nature of human γδ T cells: a biographical sketch of the highly empathetic. *Cell Mol Immunol* **10**, 21–29 (2013).
24. Rozenbaum, M. et al. Gamma-delta CAR-T cells show CAR-directed and independent activity against leukemia. *Front Immunol* **11**, (2020).
25. Hegde, M. et al. Combinational targeting offsets antigen escape and enhances effector functions of adoptively transferred T cells in glioblastoma. *Mol Ther* **21**, 2087–2101 (2013).
26. Blat, D., Zigmond, E., Alteber, Z., Waks, T. & Eshhar, Z. Suppression of murine colitis and its associated cancer by carcinoembryonic antigen-specific regulatory T cells. *Mol Ther* **22**, 1018–1028 (2014).
27. Xu, W. et al. Adoptive transfer of induced-treg cells effectively attenuates murine airway allergic inflammation. *PLoS One* **7**, e40314 (2012).
28. Mukhatayev, Z. et al. Antigen specificity enhances disease control by tregs in Vitiligo. *Front Immunol* **11**, (2020).
29. ClinicalTrials.gov. Safety and clinical activity of QEL-001 in A2-mismatch liver transplant patients (LIBERATE). *Identifier*: NCT05234190 https://clinicaltrials.gov/study/NCT05 234190 (2024).
30. ClinicalTrials.gov. Safety & tolerability study of chimeric antigen receptor T-Reg cell therapy in living donor renal transplant recipients (STEADFAST). *Identifier*: NCT04817774 https://clinicaltrials.gov/study/NCT04817774 (2024).
31. Suh, H. C., Pohl, K., Javier, A. P. L., Slamon, D. J. & Chute, J. P. Effect of dendritic cells (DC) transduced with chimeric antigen receptor (CAR) on CAR T-cell cytotoxicity. *J Clin Oncol* **35**, 144–144 (2017).
32. Mosser, D. M. & Edwards, J. P. Exploring the full spectrum of macrophage activation. *Nat Rev Immunol* **8**, 958–969 (2008).
33. Ritchie, D. et al. In vivo tracking of macrophage activated killer cells to sites of metastatic ovarian carcinoma. *Cancer Immunol Immunother* **56**, 155–163 (2006).
34. Klichinsky, M. et al. Human chimeric antigen receptor macrophages for cancer immunotherapy. *Nat Biotechnol* **38**, 947–953 (2020).
35. Levine, B. L. Performance-enhancing drugs: design and production of redirected chimeric antigen receptor (CAR) T cells. *Cancer Gene Ther* **22**, 79–84 (2015).
36. Yang, H., Hao, Y., Qi, C. Z., Chai, X. & Wu, E. Q. Estimation of total costs in pediatric and young adult patients with relapsed or refractory acute lymphoblastic leukemia receiving tisagenlecleucel from a U.S. *Hospital's Perspective J Manag Care Spec Pharm* **26**, 971–980 (2020).
37. Lee, D. W. et al. ASTCT consensus grading for cytokine release syndrome and neurologic toxicity associated with immune effector cells. *Biol Blood Marrow Transplant* **25**, 625–638 (2019).
38. Santomasso, B. D. et al. Clinical and biological correlates of neurotoxicity associated with CAR T-cell therapy in patients with B-cell acute lymphoblastic leukemia. *Cancer Discov* **8**, 958–971 (2018).
39. Brudno, J. N. & Kochenderfer, J. N. Recent advances in CAR T-cell toxicity: Mechanisms, manifestations and management. *Blood Rev* **34**, 45–55 (2019).
40. Hay, K. A. et al. Kinetics and biomarkers of severe cytokine release syndrome after CD19 chimeric antigen receptor–modified T-cell therapy. *Blood* **130**, 2295–2306 (2017).
41. Fraietta, J. A. et al. Determinants of response and resistance to CD19 chimeric antigen receptor (CAR) T cell therapy of chronic lymphocytic leukemia. *Nat Med* **24**, 563–571 (2018).

42. Philip, M. et al. Chromatin states define tumour-specific T cell dysfunction and reprogramming. *Nature* **545**, 452–456 (2017).
43. Kumar, A. R. K., Shou, Y., Chan, B., & Tay, A. Materials for improving immune cell transfection. *Adv Mater* **33**, 1–21 (2021).
44. Amini, L. et al. Preparing for CAR T cell therapy: patient selection, bridging therapies and lymphodepletion. *Nat Rev Clin Oncol* **19**, 342–355 (2022).
45. Abou-el-Enein, M. et al. Scalable manufacturing of CAR T CELLS FOR CANCER IMMUNOTHERAPY. *Blood Cancer Discov* **2**, 408–422 (2021).
46. Lanitis, E., Coukos, G. & Irving, M. All systems go: converging synthetic biology and combinatorial treatment for CAR-T cell therapy. *Curr Opin Biotechnol* **65**, 75–87 (2020).
47. Rafiq, S., Hackett, C. S. & Brentjens, R. J. Engineering strategies to overcome the current roadblocks in CAR T cell therapy. *Nat Rev Clin Oncol* **17**, 147–167 (2020).
48. Stewart, M. P. et al. In vitro and ex vivo strategies for intracellular delivery. *Nature* **538**, 183–192 (2016).
49. Tay, A. The benefits of going small: nanostructures for mammalian cell transfection. *ACS Nano* **14**, 7714–7721 (2020).
50. Moretti, A. et al. The past, present, and future of non-viral CAR T cells. *Front Immunol* **13**, (2022).
51. Kowalski, P. S., Rudra, A., Miao, L. & Anderson, D. G. Delivering the messenger: advances in technologies for therapeutic mRNA delivery. *Mol Ther* **27**, 710–728 (2019).
52. Vormittag, P., Gunn, R., Ghorashian, S. & Veraitch, F. S. A guide to manufacturing CAR T cell therapies. *Curr Opin Biotechnol* **53**, 164–181 (2018).
53. Bushman, F. D. Retroviral insertional mutagenesis in humans: evidence for four genetic mechanisms promoting expansion of cell clones. *Mol Ther* **28**, 352–356 (2020).
54. Hacein-Bey-Abina, S. et al. Efficacy of gene therapy for X-linked severe combined immunodeficiency. *N Engl J Med* **363**, 355–364 (2010).
55. Fraietta, J. A. et al. Disruption of TET2 promotes the therapeutic efficacy of CD19-targeted T cells. *Nature* **558**, 307–312 (2018).
56. Nobles, C. L. et al. CD19-targeting CAR T cell immunotherapy outcomes correlate with genomic modification by vector integration. *J Clin Investig* **130**, 673–685 (2019).
57. Nakai, H. et al. AAV serotype 2 vectors preferentially integrate into active genes in mice. *Nat Genet* **34**, 297–302 (2003).
58. Donsante, A. et al. AAV vector integration sites in mouse hepatocellular carcinoma. *Science* **317**, 477–477 (2007).
59. Cattoglio, C. et al. Hot spots of retroviral integration in human CD34+ hematopoietic cells. *Blood* **110**, 1770–1778 (2007).
60. Lamers, C. H. J. et al. Immune responses to transgene and retroviral vector in patients treated with ex vivo–engineered T cells. *Blood* **117**, 72–82 (2011).
61. Cromer, M. K. et al. Global transcriptional response to CRISPR/Cas9-AAV6-based genome editing in CD34+ hematopoietic stem and progenitor cells. *Mol Ther* **26**, 2431–2442 (2018).
62. US FDA. FDA investigating serious risk of T-cell malignancy following BCMA-directed or CD19-directed autologous Chimeric Antigen Receptor (CAR) T cell immunotherapies. https://www.fda.gov/vaccines-blood-biologics/safety-availability-biologics/fda-investigating-serious-risk-t-cell-malignancy-following-bcma-directed-or-cd19-directed-autologous (2023).
63. Ghilardi, G. et al. T cell lymphoma and secondary primary malignancy risk after commercial CAR T cell therapy. *Nat Med* **30**, 984–989 (2024).
64. Levine, B. L. et al. Unanswered questions following reports of secondary malignancies after CAR-T cell therapy. *Nat Med* **30**, 338–341 (2024).
65. Naldini, L. et al. In vivo gene delivery and stable transduction of nondividing cells by a lentiviral vector. *Science (1979)* **272**, 263–267 (1996).
66. Bulcha, J. T., Wang, Y., Ma, H., Tai, P. W. L. & Gao, G. Viral vector platforms within the gene therapy landscape. *Signal Transduct Target Ther* **6**, 53 (2021).
67. Gargett, T. & Brown, M. P. The inducible caspase-9 suicide gene system as a "safety switch" to limit on-target, off-tumor toxicities of chimeric antigen receptor T cells. *Front Pharmacol* **5**, 1–7 (2014).
68. Tay, A. & Melosh, N. Transfection with nanostructure electro-injection is minimally perturbative. *Adv Ther (Weinh)* **2**, (2019).

69. van der Loo, J. C. M. & Wright, J. F. Progress and challenges in viral vector manufacturing. *Hum Mol Genet* 25, R42–R52 (2016).

70. Weaver, J. C. & Chizmadzhev, Yu. A. Theory of electroporation: A review. *Bioelectrochem Bioenerg* 41, 135–160 (1996).

71. Almåsbak, H. et al. Transiently redirected T cells for adoptive transfer. *Cytotherapy* 13, 629–640 (2011).

72. Birkholz, K. et al. Transfer of mRNA encoding recombinant immunoreceptors reprograms CD4+ and CD8+ T cells for use in the adoptive immunotherapy of cancer. *Gene Ther* 16, 596–604 (2009).

73. Schumann, K. et al. Generation of knock-in primary human T cells using Cas9 ribonucleo-proteins. *Proc Natl Acad Sci* 112, 10437–10442 (2015).

74. Eyquem, J. et al. Targeting a CAR to the TRAC locus with CRISPR/Cas9 enhances tumour rejection. *Nature* 543, 113–117 (2017).

75. Roth, T. L. et al. Reprogramming human T cell function and specificity with non-viral genome targeting. *Nature* 559, 405–409 (2018).

76. Stenger, D. et al. Endogenous TCR promotes in vivo persistence of CD19-CAR-T cells compared to a CRISPR/Cas9-mediated TCR knockout CAR. *Blood* 136, 1407–1418 (2020).

77. Bishop, D. C. et al. CAR T cell generation by piggyBac transposition from linear Doggybone DNA vectors requires transposon DNA-flanking regions. *Mol Ther Methods Clin Dev* 17, 359–368 (2020).

78. Suematsu, M. et al. PiggyBac transposon-mediated CD19 chimeric antigen receptor-T cells derived from CD45RA-positive peripheral blood mononuclear cells possess potent and sustained antileukemic function. *Front Immunol* 13, (2022).

79. Micklethwaite, K. P. et al. Investigation of product-derived lymphoma following infusion of *piggyBac* -modified CD19 chimeric antigen receptor T cells. *Blood* 138, 1391–1405 (2021).

80. Kebriaei, P. et al. Phase I trials using sleeping beauty to generate CD19-specific CAR T cells. *J Clin Investig* 126, 3363–3376 (2016).

81. Harris, E. & Elmer, J. J. Optimization of electroporation and other non-viral gene delivery strategies for T cells. *Biotechnol Prog* 37, 1–8 (2021).

82. Pliquett, U. Joule heating during solid tissue electroporation. *Med Biol Eng Comput* 41, 215–219 (2003).

83. Ionescu-Zanetti, C., Blatz, A. & Khine, M. Electrophoresis-assisted single-cell electroporation for efficient intracellular delivery. *Biomed Microdevices* 10, 113–116 (2008).

84. Tay, A., Pavesi, A., Yazdi, S. R., Lim, C. T. & Warkiani, M. E. Advances in microfluidics in combating infectious diseases. *Biotechnol Adv* 34, 404–421 (2016).

85. Yeo, L. Y., Chang, H., Chan, P. P. Y. & Friend, J. R. Microfluidic devices for bioapplications. *Small* 7, 12–48 (2011).

86. Adamo, A. & Jensen, K. F. Microfluidic based single cell microinjection. *Lab Chip* 8, 1258 (2008).

87. Sharei, A. et al. Cell squeezing as a robust, microfluidic intracellular delivery platform. *J Vis Exp* (2013) doi:10.3791/50980

88. Hallow, D. M. et al. Shear-induced intracellular loading of cells with molecules by controlled microfluidics. *Biotechnol Bioeng* 99, 846–854 (2008).

89. Dixit, H. G. et al. Massively-parallelized, deterministic mechanoporation for intracellular delivery. *Nano Lett* 20, 860–867 (2020).

90. Uchugonova, A., Breunig, H. G., Batista, A. & König, K. Optical reprogramming of human cells in an ultrashort femtosecond laser microfluidic transfection platform. *J Biophotonics* 9, 942–947 (2016).

91. Deng, Y. et al. Intracellular Delivery of Nanomaterials via an Inertial Microfluidic Cell Hydroporator. *Nano Lett* 18, 2705–2710 (2018).

92. Jarrell, J. A. et al. Intracellular delivery of mRNA to human primary T cells with microfluidic vortex shedding. *Sci Rep* 9, 3214 (2019).

93. Park, J. C. et al. Cell Squeeze: driving more effective CD8 T-cell activation through cytosolic antigen delivery. *Immuno-Oncol Technol* 16, 100091 (2022).

94. Belling, J. N. et al. Acoustofluidic sonoporation for gene delivery to human hematopoietic stem and progenitor cells. *Proc Natl Acad Sci* 117, 10976–10982 (2020).

95. Jarrell, J. A. et al. Numerical optimization of microfluidic vortex shedding for genome editing T cells with Cas9. *Sci Rep* **11**, 11818 (2021).

96. Sevenler, D. & Toner, M. High throughput intracellular delivery by viscoelastic mechanoporation. *Nat Commun* **15**, 115 (2024).

97. Choi, Y. et al. A high throughput microelectroporation device to introduce a chimeric antigen receptor to redirect the specificity of human T cells. *Biomed Microdevices* **12**, 855–863 (2010).

98. Sharei, A. et al. A vector-free microfluidic platform for intracellular delivery. *Proc Natl Acad Sci* **110**, 2082–2087 (2013).

99. Kumar, A. et al. Mechanoporation: Toward single cell approaches. in *Handbook of Single Cell Technologies* 1–29 (Springer Singapore, Singapore, 2018). doi:10.1007/978-981-10-4857-9_3-1

100. Sharei, A. et al. Ex vivo cytosolic delivery of functional macromolecules to immune cells. *PLoS One* **10**, e0118803 (2015).

101. Li, J. et al. Microfluidic-enabled intracellular delivery of membrane impermeable inhibitors to study target engagement in human primary cells. *ACS Chem Biol* **12**, 2970–2974 (2017).

102. Ding, X. et al. High-throughput nuclear delivery and rapid expression of DNA via mechanical and electrical cell-membrane disruption. *Nat Biomed Eng* **1**, 0039 (2017).

103. Geng, T. & Lu, C. Microfluidic electroporation for cellular analysis and delivery. *Lab Chip* **13**, 3803–3821 (2013).

104. Geng, T. et al. Flow-through electroporation based on constant voltage for large-volume transfection of cells. *J Control Release* **144**, 91–100 (2010).

105. Lissandrello, C. A. et al. High-throughput continuous-flow microfluidic electroporation of mRNA into primary human T cells for applications in cellular therapy manufacturing. *Sci Rep* **10**, 18045 (2020).

106. Welch, M. et al. High-throughput CRISPR/Cas9 mediated gene editing of primary human T cells in a microfluidic device for cellular therapy manufacturing. *Adv Mater Technol* **8**, 1–12 (2023).

107. Selmeczi, D., Hansen, T. S., Met, Ö., Svane, I. M. & Larsen, N. B. Efficient large volume electroporation of dendritic cells through micrometer scale manipulation of flow in a disposable polymer chip. *Biomed Microdevices* **13**, 383–392 (2011).

108. DiTommaso, T. et al. Cell engineering with microfluidic squeezing preserves functionality of primary immune cells in vivo. *Proc Natl Acad Sci* **115**, (2018).

109. Rendall, H. A. et al. High-throughput optical injection of mammalian cells using a Bessel light beam. *Lab Chip* **12**, 4816 (2012).

110. Meacham, J. M., Durvasula, K., Degertekin, F. L. & Fedorov, A. G. Enhanced intracellular delivery via coordinated acoustically driven shear mechanoporation and electrophoretic insertion. *Sci Rep* **8**, 3727 (2018).

111. Hur, J. et al. Genetically stable and scalable nanoengineering of human primary T cells via cell mechanoporation. *Nano Lett* **23**, 7341–7349 (2023).

112. Belling, J. N. et al. Lipid-bicelle-coated microfluidics for intracellular delivery with reduced fouling. *ACS Appl Mater Interfaces* **12**, 45744–45752 (2020).

113. Kreuter, J. Nanoparticles—a historical perspective. *Int J Pharm* **331**, 1–10 (2007).

114. Chen, G., Roy, I., Yang, C. & Prasad, P. N. Nanochemistry and nanomedicine for nanoparticle-based diagnostics and therapy. *Chem Rev* **116**, 2826–2885 (2016).

115. Islam, F. et al. Exploring the journey of Zinc Oxide Nanoparticles (ZnO-NPs) toward biomedical applications. *Dent Mater* **15**, 2160 (2022).

116. Wang, E. C. & Wang, A. Z. Nanoparticles and their applications in cell and molecular biology. *Integr Biol* **6**, 9–26 (2014).

117. Nam, J. et al. Cancer nanomedicine for combination cancer immunotherapy. *Nat Rev Mater* **4**, 398–414 (2019).

118. Riley, R. S., June, C. H., Langer, R. & Mitchell, M. J. Delivery technologies for cancer immunotherapy. *Nat Rev Drug Discov* **18**, 175–196 (2019).

119. Balakrishnan, P. B. & Sweeney, E. E. Nanoparticles for enhanced adoptive T cell therapies and future perspectives for CNS tumors. *Front Immunol* **12**, (2021).

120. Mao, H.-Q. et al. Chitosan-DNA nanoparticles as gene carriers: synthesis, characterization and transfection efficiency. *J Control Release* **70**, 399–421 (2001).

121. Prabha, S., Zhou, W.-Z., Panyam, J. & Labhasetwar, V. Size-dependency of nanoparticle-mediated gene transfection: studies with fractionated nanoparticles. *Int J Pharm* **244**, 105–115 (2002).

122. Slowing, I., Viveroescoto, J., Wu, C. & Lin, V. Mesoporous silica nanoparticles as controlled release drug delivery and gene transfection carriers☆. *Adv Drug Deliv Rev* **60**, 1278–1288 (2008).

123. Adjei, I. M., Sharma, B. & Labhasetwar, V. Nanoparticles: Cellular uptake and cytotoxicity. in 73–91 (2014). doi:10.1007/978-94-017-8739-0_5

124. Iversen, T.-G., Skotland, T. & Sandvig, K. Endocytosis and intracellular transport of nanoparticles: Present knowledge and need for future studies. *Nano Today* **6**, 176–185 (2011).

125. Kovtun, A., Heumann, R. & Epple, M. Calcium phosphate nanoparticles for the transfection of cells. *Biomed Mater Eng* **19**, 241–247 (2009).

126. Billingsley, M. M. et al. Ionizable lipid nanoparticle-mediated mRNA delivery for human CAR T cell engineering. *Nano Lett* **20**, 1578–1589 (2020).

127. Yu, Q. et al. Self-assembled nanoparticles prepared from low-molecular-weight PEI and low-generation PAMAM for EGFRvIII-chimeric antigen receptor gene loading and T-cell transient modification. *Int J Nanomedicine* **15**, 483–495 (2020).

128. Parayath, N. N., Stephan, S. B., Koehne, A. L., Nelson, P. S. & Stephan, M. T. In vitro-transcribed antigen receptor mRNA nanocarriers for transient expression in circulating T cells in vivo. *Nat Commun* **11**, 6080 (2020).

129. Nakamura, T. et al. Small-sized, stable lipid nanoparticle for the efficient delivery of siRNA to human immune cell lines. *Sci Rep* **6**, 37849 (2016).

130. Adjei, I. M. et al. Functional recovery of natural killer cell activity by nanoparticle-mediated delivery of transforming growth factor beta 2 small interfering RNA. *J Interdiscip Nanomed* **4**, 98–112 (2019).

131. Xue, H. Y., Liu, S. & Wong, H. L. Nanotoxicity: a key obstacle to clinical translation of siRNA-based nanomedicine. *Nanomedicine* **9**, 295–312 (2014).

132. Smith, T. T. et al. In situ programming of leukaemia-specific T cells using synthetic DNA nanocarriers. *Nat Nanotechnol* **12**, 813–820 (2017).

133. Rurik, J. G. et al. CAR T cells produced in vivo to treat cardiac injury. *Science (1979)* **375**, 91–96 (2022).

134. Kim, K.-S. et al. Multifunctional nanoparticles for genetic engineering and bioimaging of natural killer (NK) cell therapeutics. *Biomaterials* **221**, 119418 (2019).

135. Wilk, A. J. et al. Charge-altering releasable transporters enable phenotypic manipulation of natural killer cells for cancer immunotherapy. *Blood Adv* **4**, 4244–4255 (2020).

136. Ye, Z. et al. In vitro engineering chimeric antigen receptor macrophages and T cells by lipid nanoparticle-mediated mRNA delivery. *ACS Biomater Sci Eng* **8**, 722–733 (2022).

137. Cruz, L. J. et al. Targeting nanoparticles to CD40, DEC-205 or CD11c molecules on dendritic cells for efficient CD8+ T cell response: A comparative study. *J Control Release* **192**, 209–218 (2014).

138. McKinlay, C. J., Benner, N. L., Haabeth, O. A., Waymouth, R. M. & Wender, P. A. Enhanced mRNA delivery into lymphocytes enabled by lipid-varied libraries of charge-altering releasable transporters. *Proc Natl Acad Sci* **115**, (2018).

139. Verma, M. et al. The landscape for lipid-nanoparticle-based genomic medicines. *Nat Rev Drug Discov* **22**, 349–350 (2023).

140. Kulkarni, J. A., Witzigmann, D., Chen, S., Cullis, P. R. & van der Meel, R. Lipid nanoparticle technology for clinical translation of siRNA therapeutics. *Acc Chem Res* **52**, 2435–2444 (2019).

141. Schlich, M. et al. Cytosolic delivery of nucleic acids: The case of ionizable lipid nanoparticles. *Bioeng Transl Med* **6**, (2021).

142. Sahay, G. et al. Efficiency of siRNA delivery by lipid nanoparticles is limited by endocytic recycling. *Nat Biotechnol* **31**, 653–658 (2013).

143. Gilleron, J. et al. Image-based analysis of lipid nanoparticle–mediated siRNA delivery, intracellular trafficking and endosomal escape. *Nat Biotechnol* **31**, 638–646 (2013).

144. Abildgaard, M. H., Brynjólfsdóttir, S. H. & Frankel, L. B. The autophagy–RNA interplay: degradation and beyond. *Trends Biochem Sci* **45**, 845–857 (2020).

145. Soutschek, J. et al. Therapeutic silencing of an endogenous gene by systemic administration of modified siRNAs. *Nature* 432, 173–178 (2004).
146. Karimi, M. et al. Smart micro/nanoparticles in stimulus-responsive drug/gene delivery systems. *Chem Soc Rev* 45, 1457–1501 (2016).
147. Ray, P. C., Yu, H. & Fu, P. P. Toxicity and environmental risks of nanomaterials: challenges and future needs. *J Environ Sci Health C* 27, 1–35 (2009).
148. Stater, E. P., Sonay, A. Y., Hart, C. & Grimm, J. The ancillary effects of nanoparticles and their implications for nanomedicine. *Nat Nanotechnol* 16, 1180–1194 (2021).
149. Lewinski, N., Colvin, V. & Drezek, R. Cytotoxicity of nanoparticles. *Small* 4, 26–49 (2008).
150. Xuan, L., Ju, Z., Skonieczna, M., Zhou, P. & Huang, R. Nanoparticles-induced potential toxicity on human health: Applications, toxicity mechanisms, and evaluation models. *MedComm (Beijing)* 4, 1–39 (2023).
151. Wei, X. et al. Cationic nanocarriers induce cell necrosis through impairment of Na+/K+-ATPase and cause subsequent inflammatory response. *Cell Res* 25, 237–253 (2015).
152. Mostaghaci, B., Susewind, J., Kickelbick, G., Lehr, C.-M. & Loretz, B. Transfection system of amino-functionalized calcium phosphate nanoparticles: in vitro efficacy, biodegradability, and immunogenicity study. *ACS Appl Mater Interfaces* 7, 5124–5133 (2015).
153. Przybylski, S. et al. Influence of nanoparticle-mediated transfection on proliferation of primary immune cells in vitro and in vivo. *PLoS One* 12, e0176517 (2017).
154. Xie, X. et al. Nanostraw–electroporation system for highly efficient intracellular delivery and transfection. *ACS Nano* 7, 4351–4358 (2013).
155. Cao, Y. et al. Universal intracellular biomolecule delivery with precise dosage control. *Sci Adv* 4, (2018).
156. Tay, A. & Melosh, N. Mechanical stimulation after centrifuge-free nano-electroporative transfection is efficient and maintains long-term T cell functionalities. *Small* 17, 1–11 (2021).
157. Wang, Z. et al. Interrogation of Cellular Innate Immunity by Diamond-Nanoneedle-Assisted Intracellular Molecular Fishing. *Nano Lett* 15, 7058–7063 (2015).
158. Wang, Y. et al. Poking cells for efficient vector-free intracellular delivery. *Nat Commun* 5, 4466 (2014).
159. Higgins, S. G. et al. High-aspect-ratio nanostructured surfaces as biological metamaterials. *Adv Mater* 32, 1–44 (2020).
160. Shalek, A. K. et al. Nanowire-mediated delivery enables functional interrogation of primary immune cells: application to the analysis of chronic lymphocytic leukemia. *Nano Lett* 12, 6498–6504 (2012).
161. Shokouhi, A.-R. et al. Electroactive nanoinjection platform for intracellular delivery and gene silencing. *J Nanobiotechnology* 21, 273 (2023).
162. Schmiderer, L. et al. Efficient and nontoxic biomolecule delivery to primary human hematopoietic stem cells using nanostraws. *Proc Natl Acad Sci* 117, 21267–21273 (2020).
163. Cao, Y. et al. Nontoxic nanopore electroporation for effective intracellular delivery of biological macromolecules. *Proc Natl Acad Sci* 116, 7899–7904 (2019).
164. Qu, Y., Zhang, Y., Yu, Q. & Chen, H. Surface-mediated intracellular delivery by physical membrane disruption. *ACS Appl Mater Interfaces* 12, 31054–31078 (2020).
165. Yamagishi, A. et al. Direct delivery of Cas9-sgRNA ribonucleoproteins into cells using a nanoneedle array. *Appl Sci* 9, 965 (2019).
166. Kim, H. et al. Flexible elastomer patch with vertical silicon nanoneedles for intracellular and intratissue nanoinjection of biomolecules. *Sci Adv* 4, 1–8 (2018).
167. Shalek, A. K. et al. Vertical silicon nanowires as a universal platform for delivering biomolecules into living cells. *Proc Natl Acad Sci* 107, 1870–1875 (2010).
168. Gopal, S. et al. Porous silicon nanoneedles modulate endocytosis to deliver biological payloads. *Adv Mater* 31, (2019).
169. Xie, X. et al. Mechanical model of vertical nanowire cell penetration. *Nano Lett* 13, 6002–6008 (2013).
170. Zou, J., Li, J., Chen, T. & Li, X. Penetration mechanism of cells by vertical nanostructures. *Phys Rev E* 102, 052401 (2020).
171. He, G. et al. Nanoneedle platforms: The many ways to pierce the cell membrane. *Adv Funct Mater* 30, (2020).

172. Na, Y.-R. et al. Probing enzymatic activity inside living cells using a nanowire–cell "sandwich" assay. *Nano Lett* **13**, 153–158 (2013).
173. Tieu, T., Alba, M., Elnathan, R., Cifuentes-Rius, A. & Voelcker, N. H. Advances in porous silicon–based nanomaterials for diagnostic and therapeutic applications. *Adv Ther (Weinh)* **2**, 1–25 (2019).
174. Alhmoud, H., Brodoceanu, D., Elnathan, R., Kraus, T. & Voelcker, N. H. A MACEing silicon: Towards single-step etching of defined porous nanostructures for biomedicine. *Prog Mater Sci* **116**, 100636 (2021).
175. Chiappini, C. et al. Biodegradable nanoneedles for localized delivery of nanoparticles in vivo: exploring the biointerface. *ACS Nano* **9**, 5500–5509 (2015).
176. Chiappini, C. et al. Biodegradable silicon nanoneedles delivering nucleic acids intracellularly induce localized in vivo neovascularization. *Nat Mater* **14**, 532–539 (2015).
177. Chen, Y. et al. Efficient non-viral CAR-T cell generation via silicon-nanotube-mediated transfection. *Mater Today* **63**, 8–17 (2023).
178. Chen, Y. et al. Cellular deformations induced by conical silicon nanowire arrays facilitate gene delivery. *Small* **15**, (2019).
179. Shokouhi, A. et al. Engineering efficient CAR-T cells via electroactive nanoinjection. *Adv Mater* **35**, 1–10 (2023).
180. He, G. et al. Multifunctional branched nanostraw-electroporation platform for intracellular regulation and monitoring of circulating tumor Cells. *Nano Lett* **19**, 7201–7209 (2019).
181. Choi, M. et al. Intracellular delivery of bioactive cargos to hard-to-transfect cells using carbon nanosyringe arrays under an applied centrifugal g -force. *Adv Healthc Mater* **5**, 101–107 (2016).
182. Zhang, B., Shi, Y., Miyamoto, D., Nakazawa, K. & Miyake, T. Nanostraw membrane stamping for direct delivery of molecules into adhesive cells. *Sci Rep* **9**, 6806 (2019).
183. Cai, D. et al. Highly efficient molecular delivery into mammalian cells using carbon nanotube spearing. *Nat Methods* **2**, 449–454 (2005).
184. Jiang, J. et al. Coupling of nanostraws with diverse physicochemical perforation strategies for intracellular DNA delivery. *J Nanobiotechnology* **22**, 131 (2024).
185. He, G. et al. Fabrication of various structures of nanostraw arrays and their applications in gene delivery. *Adv Mater Interfaces* **5**, 1–8 (2018).
186. Wen, R. et al. Intracellular delivery and sensing system based on electroplated conductive nanostraw arrays. *ACS Appl Mater Interfaces* **11**, 43936–43948 (2019).
187. Lard, M., Ho, B. D., Beech, J. P., Tegenfeldt, J. O. & Prinz, C. N. Use of dielectrophoresis for directing T cells to microwells before nanostraw transfection: modelling and experiments. *RSC Adv* **12**, 30295–30303 (2022).
188. Pathak, N. et al. Cellular delivery of large functional proteins and protein–nucleic acid constructs via localized electroporation. *Nano Lett* **23**, 3653–3660 (2023).
189. Liu, J. et al. Nanopore electroporation device for DNA transfection into various spreading and nonadherent cell types. *ACS Appl Mater Interfaces* **15**, 50015–50033 (2023).
190. Hu, T., Kumar, A. R., Luo, Y. & Tay, A. Automating CAR-T transfection with micro and nano-technologies. *Small Methods* (2023) doi:10.1002/smtd.202301300
191. Elnathan, R., Tay, A., Voelcker, N. H. & Chiappini, C. The start-ups taking nanoneedles into the clinic. *Nat Nanotechnol* (2022) doi:10.1038/s41565-022-01158-5

5   Bioreactor technology for CAR-T cell manufacturing

Yikai Luo and Andy Kah Ping Tay

Introduction

Cells are a source of therapeutic medicine, and bioreactors provide cells with suitable environments to proliferate. There is a close tie between the evolution of bioreactors and the use of cells. The first attempt at growing cells dates back to 2000 B.C. to preserve food in sealed clay pots. Instead of finding that the foods were well preserved, our ancestors found that the food tasted differently, leading to the discovery of fermentation. The clay pot fermenters are pivotal to the origin of bioreactors as they provide conditions for cell growth. From the 1900s onwards, scientists significantly expanded the functions of bioreactors.

At a time when most of the academia believed that it was impossible to grow yeast cells outside of their natural soil environment, Werkman and colleagues launched a series of single cell culture trials growing yeast cells using synthetic medium with medium E created by Fulmer and Nelson. They succeeded in obtaining some types of yeast cells such as *Saccharomyces cerevisiae* and *Torula rosea*, demonstrating a breakthrough that it was possible to grow yeast cells using a synthetic medium [1]. This would later lay the foundation to grow yeast cells in a controlled manner to reduce batch-to-batch differences.

In the brewery industry at that time, Siau and colleagues experienced frequent contamination. To confirm the contaminated sources, they cultured sample solutions from various parts of the fermenters in static flasks. By counting the average number of organisms across different cultured samples, they found that the contamination locations were at the coolers. Suggestions were later made by using wide-shallow coolers as well as closed-refrigeration systems to reduce the risk of contamination [2].

Although there were multiple examples of growing cells in artificial environments, they were limited to small scales. The industrialization of large-scale bioreactors only started in the mid-1950s when countries started mass vaccination campaigns to protect their citizens from infectious diseases. In the UK, the polio immunization campaign started in 1956 for all citizens aged below 40 by offering *inactivated polio vaccines (IP)* [3]. The main ingredient of the IP vaccines, the polio virus was originally cultured with primary monkey kidney cells in roller tubes [4]. This viral culture procedure was improved by replacing primary monkey kidney cells with subcultured kidney cells with three substrates with microcarriers in suspension bioreactors [5–6]. The cultured virus was chemically inactivated by formaldehyde. The suspension bioreactors with the microcarriers helped to produce a large number of IP vaccines and cater to the high vaccine demands. In the United States, attributing to the mass vaccine productions, by the late 1950s

DOI: 10.1201/9781032660752-5

diphtheria cases had been largely reduced by the immunization campaign of *diphtheria and tetanus toxoids and whole-cell pertussis vaccine (DTP)* [3]. As an example of the bioreactor used in producing DTP, the whole-cell pertussis vaccines, as one of the combined vaccines in DTP, were manufactured in many 1 L Roux bottles for bacteria growth [7]. Such vaccine campaigns would not have been successful if the manufacturing of vaccines did not meet demands. Bioreactors enabled a rapid response chain of vaccine supplies to combat preventable diseases.

Industrial bioreactors can be largely categorized into two parts: heterogeneous bioreactors where cells and growth medium are separated to create intentional gradients, and homogeneous bioreactors where every ingredient is uniformly distributed in the chamber [8]. Over time, scientists paid attention to improving both types of bioreactors.

During the vaccine campaigns, many bioreactors were designed for adherent-cell cultures. The successful industrialization of such bioreactors inspired the exploitation of the production abilities to an even larger scale. Bottle culture was one of the first attempts for large-scale adherent-cell productions. The advantages included that their individual compartments could be used for variable testing, but the downsides came from the large physical footprint required. Other inventions that came along later included the flat plate bioreactors, the microcarriers, and the packed bed bioreactors [9].

From the 1970s onwards, scientists placed greater focus on cell suspension bioreactors as they could produce higher cell density in final products. Fluidized bioreactors could mix multiple-phase liquid well with low energy consumption, and the volume of such bioreactors could go up to 1,000 L for yeast-to-ethanol cultures [10]. Modified stirred tanks adopted the improvements from stirred tanks, and the agitation was achieved by a flexible nylon sheet at the bottom. With this design, shear-sensitive cells could grow well in these tanks. The volume of these tanks could go up to 100 L for culturing rat tumor cells [10]. Hollow fiber bioreactors allowed cells to be immobilized onto the porous surface of the fiber tubes with cylindrical lumen inside. This allowed free diffusion of nutrients from medium to cells continuously through the ultrafiltration layer. The porous surface could go up to 100,000 cm². Some of the industrial products included Baboon endogenous virus, phenolics, and fructose [10].

Since the late twentieth century, scientists have been studying how to culture hard-to-grow cells with more sophisticated bioreactor culturing environments, and this marked the era of cell therapy, gene therapy, as well as tissue engineering (TE). For cell therapy, chimeric antigen receptor (CAR) T cell therapy began to treat leukemia and lymphoma successfully [11–16]. Gene therapy had also been made possible by the production of viral vectors such as retroviruses and lentiviruses through growing human embryonic kidney (HEK) cells in bioreactors [17]. For TE, such examples were tissue-engineered human cornea by collagen-based scaffolds, liver artificial organs by microcapsules, and even by hepatocyte organoids [18–20]. Figure 5.1 describes the timeline for bioreactor history [21–23].

Factors influencing bioreactor design

This section highlights various important design factors that could alter the performances of bioreactors. When choosing or designing bioreactors, each factor should be considered thoroughly for ensuring bioreactors work as expected.

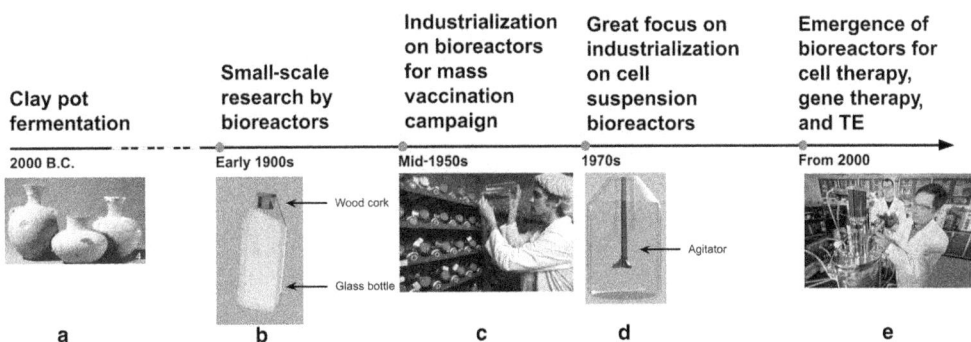

Figure 5.1 A brief outline on the history of bioreactor. A) A clay pot that was used for food fermentation during 2000 B.C., which was the first instance of bioreactor application (reproduced with permission [21], Copyright 2019, PNAS at Creative Commons licenses). B) A transparent glass container with a wood cork for cell cultures for small-scale research. The wood cork was used for gas transfers (author's 3D model interpretation). C) Roller bottles on a shelf for mass vaccination manufacturing. The shelf rotated the bottles for cells being exposed between gas and culture media (image adapted with permission [22], Wikimedia at public domain with no copyright restrictions). D) A chamber of a stirred tank with a central agitator for industrialization on cell suspension cultures. The agitator was used for better mixture of cells and culture media (author's 3D model interpretation). E) An example of a modern bioreactor that contains functions such as automative media exchange and gas feeding (image adapted with permission [23], Copyright 2015, Wikimedia at Creative Commons Attribution 2.0 Generic license).

Vessels and shear stress

The size of a bioreactor vessel controls the cell culture volume. The vessel size also correlates with the shear stress that cells will experience. Excessive amount of cell shear stress could result in cell damage, thus affecting the final product quality. It is ideal if cell shear stress limits can be quantified; therefore, precautions can be applied during cell cultures. For the example of a stir tank bioreactor, the governing equation for a cell shear stress maximum is [24]:

$$\tau = 5.33\rho(\nu\varepsilon)^{0.5} \tag{5.1}$$

where ρ is the medium density, ν is the kinematic viscosity of the medium, and ε is the power dissipated per unit mass. ε is further determined by the power equation [24]:

$$\varepsilon = \frac{P}{V_L * \rho} \tag{5.2}$$

where P is the power used by the impeller during medium agitation and V_L is the vessel volume.

It is important to study the shear stress close to the impeller region since cells experience greater forces from the impeller. The integrated shear factor (ISF) is used to express the degree of cell shear stress [24]:

$$\text{ISF} = \frac{2\pi N d}{D - d} \tag{5.3}$$

where N is the impeller speed, D is the diameter of the impeller, and d is the diameter of the vessel [24]. Equation 5.2 and equation 5.3 suggest that larger stirred tanks will result in lower cell shear stress tolerance. For example, stirred tanks are widely implemented to produce recombinant proteins for treating diseases such as diabetes, hormone deficiencies, and even viral infections. Genes of interest are delivered and integrated into the cell lines and expanded in stirred tanks. Depending on the demands for proteins of interest from the cell carriers, the size of the stirred tanks could range from only a few milliliters to 100,000 L [25]; thus, the agitation speed should be carefully calibrated for the different stirred tank sizes to ensure no shear stress damage. Computational fluid simulations could assist in predicting shear stresses. In the case of a wave bioreactor, the vessel comprises a single-used cell bag in which its polyethyl-enterephthalet material is hard enough to be considered rigid for fluid dynamic simulations [26]. The other two phases in the system are the liquid in the bag, and the air in the bag. Oftentimes, the wave bioreactor produces less shear stress compared to a stirred tank [26]. Table 5.1 summarizes the shear stresses observed with different vessel sizes and settings for wave bioreactors. All three vessel sizes produce cell expansion without significant cell viability loss.

Scalability

Bioreactors have evolved to suit different applications, and working volume is one of the major variants between bioreactors. Smaller working volumes could be as low as micro-L scale, but there are challenges in miniaturizing essential components. Larger working volumes could go up to as high as kilo-L scale, but with the risks of high shear stress and insufficient oxygen delivery [27]. Desired working volumes of bioreactors are determined by product demands. For large-demand cell products such as vaccines and insulins, easy scale-up bioreactors are suitable in these cases to implement mass productions. Allogeneic cell production targets larger populations instead of at the individual level. Mass production by scaling up ensures sufficient supplies. Cell expansion times would not be a

Table 5.1 Shear stresses with different vessel sizes and settings for wave bioreactors

Vessel sizes (L)	Rotation angles (degree)	Shaking speeds (rpm)	Shear stress (Pa)	References
2	7	16	<0.01	[13.3]
5	4,5.5,7	15,22,30	0.01–0.06	[13.2]
	In total nine combinations of different rotation angles and shaking speeds, i.e., 4 degrees with 15 rpm, 4 degrees with 22 rpm, etc.			
20	8	16	<0.01	[13.3]

significant concern since multiple pipelines can produce supply reservoirs to offset expansion waiting times. For low-demand products such as specific cell dosages for individuals, scale-out bioreactors are needed in this case for parallel manufacturing. Autologous cell sources are typically associated with low-demand products. They have an expansion cycle of two to three weeks before cells reach clinically desirable cell counts. For example, CD19-specific CAR-T cells are proven to be effective in treating malignant B cells as these B cells carry CD19 signaling pathways for being targeted [28]. CAR-T cell therapy takes blood samples from patients for leukapheresis to obtain autologous T cells as starting materials. As autologous cells from different patients are not compatible, only one type of autologous cells can be cultured in a bioreactor each time; thus, scale-out models are needed for accommodating more patients at one time and reducing dosage preparation times for a single patient by multiple pipelines. Shorter expansion times of CAR-T cells would result in higher chances of infusion within time windows.

Good manufacturing practices

New bioreactors are likely to be categorized under Class I medical devices with premarket notification requirements by the US Food and Drug Administration as they have a long history of use in cell manufacturing [29]. Good manufacturing practices (GMPs) are required for bioreactor designs under Code of Federal Regulations Title 21, Part 820 [29]. GMPs require devices' specifications to be fully characterized, and when deviations are expected, clear instructions shall be made to ensure such deviations will not compromise final product qualities. In terms of bioreactor manufacturing, cell mounting methods such as process volumes and minimum output volumes are important specifications associated with product qualities and need to be fully characterized [30]. GMP also requires a process to eliminate contamination. In bioreactor designs, using closed system pipelines is one of the most viable methods to ensure no contamination. For example, disposable bioreactors have become a new trend for recombinant protein production from suspension cells. The disposable bioreactors would enhance the closed system concept and reduce contamination risk. Otherwise, higher costs on environmental controls are needed to implement clean room facilities to avoid contamination.

Oxygen transfer rate

Oxygen is a key ingredient for cell growth. Unfortunately, only little oxygen diffuses into the liquid phase. Oxygen solubility in water is measured to be at only 8 mg/L at room temperature and is even lower in most culture media [31]. Moreover, passive oxygen diffusion in culture media is unable to support cell growth [31]. This poses challenges for cell growth in bioreactors, and thus, it is important to implement apparatus to allow sufficient oxygen diffusion during cell expansions. Oxygen transfer rate (OTR) is a quantitative measure of the rate of oxygen diffusion in a medium [31]:

$$OTR = K_L a \left(C_{sat} - C \right) \tag{5.4}$$

where $K_L a$ is the overall volumetric oxygen transfer coefficient, C_{sat} is the oxygen concentration in the liquid, and C is the bulk oxygen concentration in the liquid. The derivation of equation 5.4 suggests that the resistance for oxygen to transfer at the liquid stagnant film is the most significant under the two-film theory; thus, reducing the liquid stagnant film could increase OTR [31–32]. In bioreactors, oxygen is supplied by tubes into the liquid phase, and air bubbles are shredded by liquid agitation to reduce the size of liquid stagnant film for higher OTR. Faster agitation speed results in higher OTR; however, this produces higher shear stress that could damage cells. Thus, it is important not to exceed the cell shear stress limit for agitation speed in bioreactors by means of implementing correct geometrical and operational characteristics such as agitation methods and fluid characteristics [32]. Marin De and Florian argued that with orbital oxygen mass transfer by higher oxygen content in the chamber, a disposable orbital shaking bioreactor would result in a cheaper cost compared to the conventional stirred tanks in producing proteins [33].

Other considerations and release tests

Cell culture in-process parameters are important channels for manufacturing monitoring. Parameters such as pH, temperature, bioreactor vessel pressure, and dissolved oxygen should be measured during cell cultures [34–35]. Growth factor concentrations such as glucose and waste product concentrations such as lactate and ammonia should be monitored as well [36]. Sensors should be implemented for obtaining parameters from bioreactors. Sensors are important for monitoring the cell environment inside the bioreactor. It is also critical to release the by-products through air valves and keep the fermentation process aseptic. Sensors can be used as a prompt for by-product release once the by-product reading hits the preset threshold. Usually, such sensors are sampling ports that interfere with media in bioreactors. These invasive sensors pose risks for cell contamination. The examples for the sampling ports include the peristaltic pump with tubes that could be improved by using the thermoplastic tubing and the one-way valve syringe system [37]. Recently, scientists have been paying more attention to the sensors for single-use bioreactors. Such sensors are disposable and relatively low cost compared to the sampling ports. The disposable sensors do not impose contamination problems; instead, they are added into media in bioreactors. However, this introduces another challenge to isolate disposable sensors from final products. The examples for the disposable sensors include electrochemical sensors, 'PATsule' capsule sensors, and ion-selective field effect transistors [37–39]. Quality controls are important prior to product release. Cell counts can be obtained by machines such as the Cedex HiRes system, the Luna, and the Cellometer [40]. For CAR-T cells, the minimal cell viability should be between 70% and 80%, and the minimal cell counts should be at 1.0×10^9 [41–42]. Cell types and cell subsets can be determined by the fluorescence-activated cell sorting technique, dual-fluorescence automated cell counting, and the Trypan blue exclusion [42–43]. Cytotoxicity and cytokine profiles can be obtained from the micro-engraving immune cell arrays and the real-time antibody-modified surface plasmon resonance [44].

CAR-T cell manufacturing with bioreactors

This section introduces different bioreactors used for CAR-T cell manufacturing. Table 5.2 summarizes the culture parameters between these bioreactors for CAR-T cell cultures. Table 5.3 summarizes the feature comparisons between the typical bioreactors for CAR-T cell cultures [45–48]. Figure 5.2 shows the bioreactors described in this section [49–52]. In general, CAR-T cell manufacturing involves the types of bioreactors such as stirred tanks, gas-permeable chambers, centrifugation devices, microfluidic devices, fully automated cartridges, and wave motion devices.

For stirred tanks, an impeller with specific cutting angles facilitates the agitation of the culture media. In a stirred tank, it is possible to implement multiple impellers for a better mixture, though it is important not to cause significant cell perturbations. The cell culture performance largely depends on the agitation speed. For the culture of CAR-T cells in unbaffled Ambr 250 with a single impeller, the lower the agitation speed, the lower the cell counts. With an agitation speed of 300 rpm, the total CAR-T viable cell counts can reach 5.07×10^6 mL^{-1} [53]. Another CAR-T cell culture with a larger stirred tank can reach total CAR-T viable cell counts at 6.40×10^6 mL^{-1} [54].

An example of a bioreactor for CAR-T culture in gas-permeable chambers is the G-Rex in plates and in closed-system chambers. For the cultures in the plates, the expansion fold was observed to be 42-fold compared to the pre-culture [55]. If the culture media volume is optimized at 10 mL cm^{-2}, the expansion fold increases to 80-fold [55]. There is a linear relation between the seeding area and the final cell counts, given the same starting culture conditions, which are usually determined by the small-scale G-Rex cultures in plates. A scaled-up CAR-T cell culture using the predetermined starting conditions from the plates yields a total cell count of 3×10^9 cells for $1,000$ cm^2 [56].

For the centrifugation devices, an excellent example is the CliniMACS Prodigy system. A chamber controls the cell culture process. Before the mark of 150 mL, the cell culture process remains in the static mode and thereafter the agitation mode. During media exchange, centrifugation takes place to separate the used media and the cultured cells, followed by the fresh media addition. For a span of 12 days, the device reported provided a cell expansion of 47-fold [57].

On the contrary, the large-scale one-stop solution for CAR-T cell manufacturing is the cartridge model. A robotic arm is positioned inside a fully enclosed room-sized chamber to move the cartridge to the different stations of the manufacturing process. Cellares is a company that offers the cartridge-type bioreactor to promote the integrated development and manufacturing structure. The device can reach a CAR-T cell count of 4.8×10^8 for each cartridge over five days [58]. Other similar products include the Lonza Cocoon cartridge bioreactor and the MultiplyLabs cartridges [59–60].

For wave motion devices, the typical example is from Xuri. A rotor enables the wave motion for the cell bag that is placed on top of the plate. The motion enhances the cell growth and the nutrient exchange. The speed and the maximum angle of the wave motion are adjustable for the different volumes of the cell bags. The device can reach a CAR-T cell count of 2.0×10^{10} [61].

While Table 5.2 lists only one volume variant for each bioreactor type (if a volume is specified), many of these bioreactor types have different volume variants. It is ideal when a suitable volume variant is chosen to optimize the cost efficiency. CAR-T cell therapy products should have a minimal cell count of 1.0×10^9. For the bioreactors that can reach this cell count, the volume variant should be kept as low as possible. Otherwise, a larger volume variant needs to be used for meeting the minimal cell count for patient infusion.

Table 5.2 The culture parameters between the bioreactors for CAR-T cell cultures

Bioreactor	Control	Source	Agitation speed	CAR-T viable cell rate	Total cell count
Stirred tank (single-impeller unbaffled Ambr 250, Sartorius, Germany)	Static flask	Autologous activated T cells from three donors	From 100 to 500 rpm	> 90% for all rpm	Highest yield observed at 300 rpm at unspecified power for 7 days: 5.07×10^6 cells mL^{-1} (25-fold) for 250 mL (significantly increased compared to control)
Stirred tank (single-impeller unbaffled, customized)	Spinner flask	Autologous activated T cells from one donor	Unspecified	94%	At unspecified power for 4 days: 6.40×10^6 cells mL^{-1} for 800 mL (significantly increased compared to control)
Gas-permeable chamber (G-Rex plates, Wilson Wolf, USA)	Static culture bag	Autologous activated T cells from two donors	N.A.	Unspecified	Highest yield observed at a seeding density of 0.06×10^6 cells cm^{-2} and unspecified medium density for up to 21 days: 2.6×10^7 cells (42-fold) per well (10 cm^2)
Gas-permeable chamber (G-Rex plates, Wilson Wolf, USA)	Static culture bag	Autologous activated T cells from three donors	N.A.	Unspecified	Highest yield observed at a seeding density of 0.5×10^6 cells cm^{-2} and medium density of 10 mL cm^{-2} for up to 21 days: 4.2×10^8 cells (~ 80-fold) per well (10 cm^2)
Gas-permeable chamber (G-rex100M-CS, Wilson Wolf, USA)	Unspecified	Autologous activated T cells	N.A.	Unspecified	Highest yield observed at a seeding density of 0.5×10^6 cells cm^{-2} and medium density of 10 mL cm^{-2} for 9–10 days: 3×10^9 cells (1,000 cm^2)
Centrifugation device (CliniMACS Prodigy, Miltenyi, Germany)	Unspecified	Autologous activated T cells from three donors	Unspecified	> 96%	For 9 days: 2.4 to 2.7×10^9 (24 to 27-fold) at unspecified mL (significant increased compared to pre-culture);For 12 days: 4.7×10^9 (47-fold) at unspecified volume (significant increase compared to pre-culture)
Cartridge (Cell shuttle, Cellares, USA)	Unspecified competitor	Autologous activated T cells	N.A.	> 80%	Unspecified (final viable CAR-T cell count: ~ 4.8×10^8 cells for 5 days)
Wave motion device(Xuri, Cytiva, USA)	Unspecified	Autologous T cells	N.A.	Unspecified	For 14 days: 2.0×10^{10} for 1 L

Cell type analysis

CD4+/CD8+	CD8+ naïve type	CD8+ central memory type	CD8+ effector memory type	CD8+ terminally differentiated type	CAR receptor
1:1 for all rpm (comparable with control)	< 5% for all rpm	Lower than pre-culture (39.6%) for all rpm (comparable with control)	Comparable with control	< 5% for all rpm	Comparable with control
Unspecified	Unspecified	Unspecified	Unspecified	Unspecified	Unspecified
Unspecified	Unspecified	Unspecified	Unspecified	Unspecified	Unspecified
Unspecified	Unspecified	Unspecified	Unspecified	Unspecified	Unspecified
Unspecified	Unspecified	Unspecified	Unspecified	Unspecified	Unspecified
Unspecified	Lower than pre-culture	Higher than pre-culture	Higher than pre-culture	Lower than pre-culture	Unspecified
Two (comparable with pre-culture)	Unspecified	Unspecified	Unspecified	Unspecified	Unspecified
Unspecified	Unspecified	Unspecified	Unspecified	Unspecified	Unspecified

Table 5.3 The feature comparisons between the typical bioreactors for CAR-T cell cultures

Bioreactor	Scalability	Shear stress	Reusability	Cost	CAR-T cell yield	Other factors
Stirred tanks	Easy to scale up	High	Single use	Moderate	CAR-T cell expansions are comparable to static culture control	+ High availability; + High level of aeration; - Shear stress is sensitive to impeller shapes
Gas-permeable chambers (G-rex, Wilson Wolf, USA)	Easy to scale up	Low	Single use	Cheap	CAR-T cell expansions are comparable to static culture control: • CAR-T cell expansions are up to 20-fold increase compared to static culture control; • CAR-T cell expansions are about 4-fold decrease in manpower time compared to static culture control	+ Large culture media requires no media change during cultures; + Low seeding requirements;+ Various seeding areas: 5, 100, 500 cm^2; + Glucose concentration monitoring is viable; + Only require standard laboratory equipment: laminar flow and incubator; - Difficult to implement other in-process sample monitoring, yet rigorous sampling is needed for process control; - High risk of contamination
Centrifugation devices (CliniMACS Prodigy, Miltenyi, Germany)	Hard to scale up; hard to scale out	Low	Single use	High	CAR-T cells expansion fold compared to static control: • comparable	+ Reduced manual operation time to 6 hours 20 minutes for one culture cycle; - Requires high-cost consumables, for example, connectors, tubing - Cell culture happens in a centrifuge chamber that could introduce unpredictable culture environments
Wave motion devices	Easy to scale up	Low	Single use	Moderate	CAR-T cells expansions are lower to static culture control: • 247–1,340 fold for static culture control and 200–800 fold for wave motion devices	+ Easy to implement sensors for pH, dO_2, and temperature; + Aseptic media exchange and sampling process;+ Various cell bag volume: 100 mL to 300 L (Biostat RM, Sartorius, Germany), 300 mL to 25 L (Xuri, Cytiva, USA); - 24 hours of manual operation time with a total expansion process taking up to 10 days for one culture cycle

+ represents favorable factors, − represents unfavorable factors.

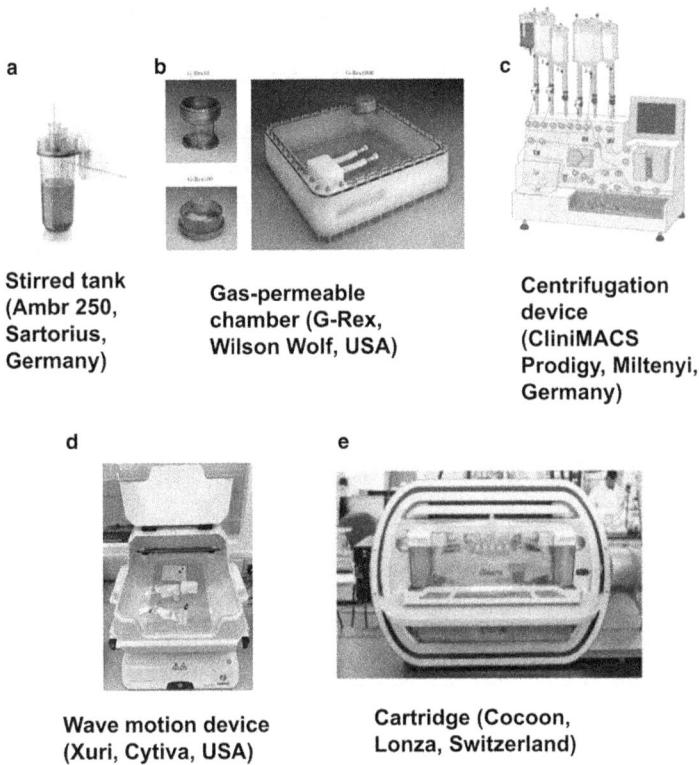

Stirred tank (Ambr 250, Sartorius, Germany)

Gas-permeable chamber (G-Rex, Wilson Wolf, USA)

Centrifugation device (CliniMACS Prodigy, Miltenyi, Germany)

Wave motion device (Xuri, Cytiva, USA)

Cartridge (Cocoon, Lonza, Switzerland)

Figure 5.2 The various bioreactors for CAR-T cell manufacturing. (a) An Ambr 250 stirred tank. This stirred tank could vary between different numbers of agitators and impellers (image adapted with permission [49], Copyright 2021, Springer Nature at Creative Commons licenses). (b) G-Rex gas-permeable chambers with different seeding areas. CAR-T cells are seeded at the bottom surface for expansion (image adapted with permission [50], Copyright 2011, Hindawi at Creative Commons Attribution License). (c) A CliniMACS centrifugation device. CAR-T cells are expanded in the centrifugation chamber with culture media and gas supplies (image adapted with permission [51], Copyright 2023, GlaxoSmithKline at Creative Commons Attribution License). (d) A Xuri wave motion device. The device introduces consistent wave motions to a cell bag for CAR-T cell expansion (author's image). (e) A Cocoon cartridge. The cartridge has a bench-top size and enables automative CAR-T cell expansion. Image adapted with permission [52], Copyright 2021, MDPI at Creative Common CC BY license.

Conclusion

A bioreactor is a key enabling technology for cell manufacturing. The ongoing trend of modern bioreactors for CAR-T manufacturing is a fully enclosed and automated system to reduce the risk of contamination, provide low batch-to-batch variations, and decrease manpower requirements. The automation of bioreactors in particular would improve current CAR-T cell manufacturing by offering decentralized manufacturing options (DMOs) to therapy providers. In a decentralized setting, each hospital or a cluster of nearby hospitals are equipped with automated CAR-T cell manufacturing facilities; thus, no centralized manufacturing locations are needed, for example, big pharmaceutical factories [62]. DMOs provide better access to patients. This is because DMOs would eliminate the needs for blood cryopreservation, thereby improving the CAR-T cell phenotypes [62]. DMOs would also provide scale-out channels for manufacturing different patients' CAR-T cells simultaneously and automatically to shorten vein-to-vein times. It is important to implement automation-viable manufacturing procedures such as adopting non-viral T cell transfection techniques [63–66] so that all CAR-T cell manufacturing processes can be included in one automated process. On another aspect, the incorporation of artificial intelligence (AI) would help identify optimal CAR-T manufacturing parameters, such as the CD4+/CD8+ ratio that relates to CAR-T cell therapeutic abilities and selected proportion of CD3+ T cells populations [63]. AI is also capable of obtaining in-process manufacturing data from sensors to control the manufacturing process [64]. In conclusion, it is incredible to see the iterations of bioreactors from the ancient times to the modern era, and continual improvements would enable cheaper and more timely CAR-T cell manufacturing for patients.

References

[1] C. H. Werkman, 'Continuous Reproduction of Micro-Organisms in Synthetic Media', *Science*, vol. 62, no. 1596, pp. 115–116, Jul. 1925, doi: 10.1126/science.62.1596.115

[2] R. L. Siau, 'Brewery Infection and Pure Yeast', *Journal of the Institute of Brewing*, vol. 12, no. 2, pp. 118–143, Mar. 1906, doi: 10.1002/j.2050-0416.1906.tb02157.x

[3] R. W. Sutter and C. Maher, 'Mass Vaccination Campaigns for Polio Eradication: An Essential Strategy for Success', in *Mass Vaccination: Global Aspects — Progress and Obstacles*, S. A. Plotkin, Ed., Berlin, Heidelberg: Springer, 2006, pp. 195–220. doi: 10.1007/3-540-36583-4_11

[4] J. E. Salk, B. L. Bennett, L. J. Lewis, E. N. Ward, and J. S. Youngner, 'Studies in Human Subjects On Active Immunization Against Poliomyelitis: 1. A Preliminary Report of Experiments in Progress', *Journal of the American Medical Association*, vol. 151, no. 13, pp. 1081–1098, Mar. 1953, doi: 10.1001/jama.1953.13.1081

[5] van Wezel, A. L., 'Growth of Cell-strains and Primary Cells on Micro-carriers in Homogeneous Culture,' *Nature*, vol. 216, no. 5110, Art. no. 5110, Oct. 1967, doi: 10.1038/216064a0

[6] A. L. van Wezel, G. van Steenis, P. van der Marel, and A. D. M. E. Osterhaus, 'Inactivated Poliovirus Vaccine: Current Production Methods and New Developments', *Reviews of Infectious Diseases*, vol. 6, pp. S335–S340, 1984.

[7] S. M. Cohen and M. W. Wheeler, 'Pertussis Vaccine Prepared with Phase-I Cultures Grown in Fluid Medium', *American Journal of Public Health and the Nation's Health*, vol. 36, no. 4, pp. 371–376, Apr. 1946.

[8] K. Alvi, 'Cell Culture Technology for Pharmaceutical and Cell-Based Therapies Edited by S. S. Ozturk and W.-S. Hu (Centocor, Inc. and University of Minnesota, respectively). CRC Press/Taylor & Francis, Boca Raton. 2006. xiv + 755 pp. 7 × 10 in. $179.95. ISBN 0-8247-5334-8', *Journal of Natural Products*, vol. 70, no. 4, pp. 712–713, Apr. 2007, doi: 10.1021/np078140a

[9] R. E. Spier and A. Kadouri, 'The Evolution of Processes for the Commercial Exploitation of Anchorage-Dependent Animal Cells', *Enzyme and Microbial Technology*, vol. 21, no. 1, pp. 2–8, Jul. 1997, doi: 10.1016/S0141-0229(96)00213-X

[10] A. Margaritis and J. B. Wallace, 'Novel Bioreactor Systems and Their Applications,' *Nature Biotechnology*, vol. 2, no. 5, Art. no. 5, May 1984, doi: 10.1038/nbt0584-447

[11] D. L. Porter, B. L. Levine, M. Kalos, A. Bagg, and C. H. June, 'Chimeric Antigen Receptor–Modified T Cells in Chronic Lymphoid Leukemia', *New England Journal of Medicine*, vol. 365, no. 8, pp. 725–733, Aug. 2011, doi: 10.1056/NEJMoa1103849

[12] S. A. Grupp et al., 'Chimeric Antigen Receptor–Modified T Cells for Acute Lymphoid Leukemia', *New England Journal of Medicine*, vol. 368, no. 16, pp. 1509–1518, Apr. 2013, doi: 10.1056/NEJMoa1215134

[13] 'Emily Whitehead', Cancer Research Institute. Accessed: Mar. 14, 2024. [Online]. Available: https://www.cancerresearch.org/stories/patients/emily-whitehead

[14] R. J. Brentjens et al., 'CD19-Targeted T Cells Rapidly Induce Molecular Remissions in Adults with Chemotherapy-Refractory Acute Lymphoblastic Leukemia', *Science Translational Medicine*, vol. 5, no. 177, pp. 177ra38, Mar. 2013, doi: 10.1126/scitranslmed.3005930

[15] J. N. Kochenderfer et al., 'Chemotherapy-Refractory Diffuse Large B-Cell Lymphoma and Indolent B-Cell Malignancies Can Be Effectively Treated With Autologous T Cells Expressing an Anti-CD19 Chimeric Antigen Receptor', *JCO*, vol. 33, no. 6, pp. 540–549, Feb. 2015, doi: 10.1200/JCO.2014.56.2025

[16] J. N. Kochenderfer et al., 'B-cell Depletion and Remissions of Malignancy Along with Cytokine-Associated Toxicity in a Clinical Trial of Anti-CD19 Chimeric-Antigen-Receptor–Transduced T cells', *Blood*, vol. 119, no. 12, pp. 2709–2720, Mar. 2012, doi: 10.1182/blood-2011-10-384388

[17] I. M. Verma and M. D. Weitzman, 'GENE THERAPY: Twenty-First Century Medicine', *Annual Review of Biochemistry*, vol. 74, no. 1, pp. 711–738, 2005, doi: 10.1146/annurev.biochem.74.050304.091637

[18] C. J. Doillon et al., 'A Collagen-Based Scaffold for a Tissue Engineered Human Cornea: Physical and Physiological Properties', *The International Journal of Artificial Organs*, vol. 26, no. 8, pp. 764–773, Aug. 2003, doi: 10.1177/039139880302600810

[19] H. W. Matthew, S. O. Salley, W. D. Peterson, and M. D. Klein, 'Complex Coacervate Microcapsules for Mammalian Cell Culture and Artificial Organ Development,' *Biotechnology Progress*, vol. 9, no. 5, pp. 510–519, 1993, doi: 10.1021/bp00023a010

[20] K. Funatsu, H. Ijima, K. Nakazawa, Y. Yamashita, M. Shimada, and K. Sugimachi, 'Hybrid Artificial Liver Using Hepatocyte Organoid Culture', *Artificial Organs*, vol. 25, no. 3, pp. 194–200, 2001, doi: 10.1046/j.1525-1594.2001.025003194.x

[21] L. Liu et al., 'The Origins of Specialized Pottery and Diverse Alcohol Fermentation Techniques in Early Neolithic China', *Proceedings of the National Academy of Sciences*, vol. 116, no. 26, pp. 12767–12774, Jun. 2019, doi: 10.1073/pnas.1902668116

[22] L. Bartlett, 1980. Accessed: Mar. 14, 2024. [Online]. Available: https://commons.wikimedia.org/wiki/File:Monoclonal_antibodies_(3).jpg

[23] USDA, *English: A bioreactor used to ferment ethanol from corncob waste being loaded with yeast.* 2013. Accessed: Mar. 14, 2024. [Online]. Available: https://commons.wikimedia.org/wiki/File:Loading_bioreactor.jpg

[24] S. Khodabakhshaghdam, A. B. Khoshfetrat, and R. Rahbarghazi, 'Alginate-chitosan Core-Shell Microcapsule Cultures of Hepatic Cells in a Small Scale Stirred Bioreactor: Impact of Shear Forces and Microcapsule Core Composition', *Journal of Biological Engineering*, vol. 15, no. 1, p. 14, Apr. 2021, doi: 10.1186/s13036-021-00265-6

[25] J. Puetz and F. M. Wurm, 'Recombinant Proteins for Industrial versus Pharmaceutical Purposes: A Review of Process and Pricing', *PRO*, vol. 7, no. 8, Art. no. 8, Aug. 2019, doi: 10.3390/pr7080476

[26] C. Zhan, E. Hagrot, L. Brandt, and V. Chotteau, 'Study of Hydrodynamics in Wave Bioreactors by Computational Fluid Dynamics Reveals a Resonance Phenomenon', *Chemical Engineering Science*, vol. 193, pp. 53–65, Jan. 2019, doi: 10.1016/j.ces.2018.08.017

[27] S. Junne and P. Neubauer, 'How Scalable and Suitable are Single-Use Bioreactors?,' *Current Opinion in Biotechnology*, vol. 53, pp. 240–247, Oct. 2018, doi: 10.1016/j.copbio.2018.04.003

[28] M. C. Milone et al., 'Engineering-Enhanced CAR T Cells for Improved Cancer Therapy', *Nature Cancer*, vol. 2, no. 8, Art. no. 8, Aug. 2021, doi: 10.1038/s43018-021-00241-5

[29] D. Lim et al., 'Bioreactor Design and Validation for Manufacturing Strategies in Tissue Engineering', *Bio-Design and Manufacturing*, vol. 5, no. 1, pp. 43–63, Jan. 2022, doi: 10.1007/s42242-021-00154-3

[30] A. Li et al., 'Advances in Automated Cell Washing and Concentration', *Cytotherapy*, vol. 23, no. 9, pp. 774–786, Sep. 2021, doi: 10.1016/j.jcyt.2021.04.003

[31] K. G. Clarke, '8 - The Oxygen Transfer Rate and Overall Volumetric Oxygen Transfer Coefficient', in *Bioprocess Engineering*, K. G. Clarke, Ed., Woodhead Publishing, 2013, pp. 147–170. doi: 10.1533/9781782421689.147

[32] V. C. Srivastava, I. M. Mishra, and S. Suresh, '2.69 - Oxygen Mass Transfer in Bioreactors.' in *Comprehensive Biotechnology (Second Edition)*, M. Moo-Young, Ed., Burlington: Academic Press, 2011, pp. 947–956. doi: 10.1016/B978-0-08-088504-9.00409-8

[33] M. De Jesus and F. M. Wurm, 'Manufacturing Recombinant Proteins in Kg-ton Quantities Using Animal Cells in Bioreactors', *European Journal of Pharmaceutics and Biopharmaceutics*, vol. 78, no. 2, pp. 184–188, Jun. 2011, doi: 10.1016/j.ejpb.2011.01.005

[34] E. Trummer et al., 'Process Parameter Shifting: Part I. Effect of DOT, pH, and Temperature on the Performance of Epo-Fc Expressing CHO Cells Cultivated in Controlled Batch Bioreactors', *Biotechnology and Bioengineering*, vol. 94, no. 6, pp. 1033–1044, 2006, doi: 10.1002/bit.21013

[35] M. G. Vander Heiden, D. R. Plas, J. C. Rathmell, C. J. Fox, M. H. Harris, and C. B. Thompson, 'Growth Factors Can Influence Cell Growth and Survival through Effects on Glucose Metabolism', *Molecular and Cellular Biology*, vol. 21, no. 17, pp. 5899–5912, Sep. 2001, doi: 10.1128/MCB.21.17.5899-5912.2001

[36] H. J. Cruz, C. M. Freitas, P. M. Alves, J. L. Moreira, and M. J. T. Carrondo, 'Effects of Ammonia and Lactate on Growth, Metabolism, and Productivity of BHK cells', *Enzyme and Microbial Technology*, vol. 27, no. 1, pp. 43–52, Jul. 2000, doi: 10.1016/S0141-0229 (00)00151-4

[37] A. Glindkamp, D. Riechers, C. Rehbock, B. Hitzmann, T. Scheper, and K. F. Reardon, 'Sensors in Disposable Bioreactors Status and Trends', in *Disposable Bioreactors*, R. Eibl and D. Eibl, Eds., Berlin, Heidelberg: Springer, 2010, pp. 145–169. doi: 10.1007/10_2009_10

[38] A. Glindkamp, D. Riechers, C. Rehbock, B. Hitzmann, T. Scheper, and K. F. Reardon, 'Sensors in Disposable Bioreactors Status and Trends', in *Disposable Bioreactors*, R. Eibl and D. Eibl, Eds., Berlin, Heidelberg: Springer, 2010, pp. 145–169. doi: 10.1007/10_2009_10

[39] P. O'Mara, A. Farrell, J. Bones, and K. Twomey, 'Staying Alive! Sensors Used for Monitoring Cell Health in Bioreactors,' *Talanta*, vol. 176, pp. 130–139, Jan. 2018, doi: 10.1016/j. talanta.2017.07.088

[40] D. Cadena-Herrera et al., 'Validation of Three Viable-Cell Counting Methods: Manual, Semi-Automated, and Automated,' *Biotechnology Reports*, vol. 7, pp. 9–16, Sep. 2015, doi: 10.1016/j.btre.2015.04.004

[41] K. A. Jalowiec et al., 'How to Collect the Minimum-Targeted CD3+ Cells for CAR-T Therapy-the Bern Approach', *Blood*, vol. 134, no. Supplement_1, p. 2457, Nov. 2019, doi: 10.1182/ blood-2019-126534

[42] E. A. Chong et al., 'CAR T Cell Viability Release Testing and Clinical Outcomes: Is There a Lower Limit?', *Blood*, vol. 134, no. 21, pp. 1873–1875, Nov. 2019, doi: 10.1182/ blood.2019002258

[43] J. E. Eyles, S. Vessillier, A. Jones, G. Stacey, C. K. Schneider, and J. Price, 'Cell Therapy Products: Focus On Issues with Manufacturing and Quality Control of Chimeric Antigen Receptor T-cell Therapies', *Journal of Chemical Technology & Biotechnology*, vol. 94, no. 4, pp. 1008–1016, 2019, doi: 10.1002/jctb.5829

[44] A. Revzin, E. Maverakis, and H.-C. Chang, 'Biosensors for Immune Cell Analysis—a Perspective,' *Biomicrofluidics*, vol. 6, no. 2, p. 021301, Apr. 2012, doi: 10.1063/1.4706845

[45] K. Swiech, K. C. R. Malmegrim, and V. Picanço-Castro, Eds., *Chimeric Antigen Receptor T Cells: Development and Production*, vol. 2086. in Methods in Molecular Biology, vol. 2086. New York, NY: Springer US, 2020. doi: 10.1007/978-1-0716-0146-4

[46] I. Ganeeva et al., 'Recent Advances in the Development of Bioreactors for Manufacturing of Adoptive Cell Immunotherapies', *Bioengineering*, vol. 9, no. 12, Art. no. 12, Dec. 2022, doi: 10.3390/bioengineering9120808

[47] M. Abou-el-Enein et al., 'Scalable Manufacturing of CAR T Cells for Cancer Immunotherapy', *Blood Cancer Discovery*, vol. 2, no. 5, pp. 408–422, Sep. 2021, doi: 10.1158/2643-3230. BCD-21-0084

[48] M. Stephenson and W. L. Grayson, 'Recent Advances in Bioreactors for Cell-Based Therapies.' *F1000Research*, Apr. 30, 2018. doi: 10.12688/f1000research.12533.1

[49] M. Rotondi et al., 'Design and Development of a New ambr250® Bioreactor Vessel for Improved Cell and Gene Therapy Applications', *Biotechnology Letters*, vol. 43, no. 5, pp. 1103–1116, May 2021, doi: 10.1007/s10529-021-03076-3

[50] N. Lapteva and J. F. Vera, 'Optimization Manufacture of Virus- and Tumor-Specific T Cells', *Stem Cells International*, vol. 2011, p. e434392, Sep. 2011, doi: 10.4061/2011/434392

[51] N. Francis et al., 'Development of an Automated Manufacturing Process for Large-scale Production of Autologous T Cell Therapies', *Molecular Therapy - Methods & Clinical Development*, vol. 31, p. 101114, Dec. 2023, doi: 10.1016/j.omtm.2023.101114

[52] A. Titov et al., 'Adoptive Immunotherapy beyond CAR T-Cells', *Cancers*, vol. 13, no. 4, Art. no. 4, Jan. 2021, doi: 10.3390/cancers13040743

[53] E. Costariol et al., 'Demonstrating the Manufacture of Human CAR-T Cells in an Automated Stirred-Tank Bioreactor', *Biotechnology Journal*, vol. 15, no. 9, p. 2000177, 2020, doi: 10.1002/biot.202000177

[54] J. Ou et al., 'Novel Biomanufacturing Platform for Large-Scale and High-Quality Human T Cells Production', *Journal of Biological Engineering*, vol. 13, no. 1, p. 34, Apr. 2019, doi: 10.1186/s13036-019-0167-2

[55] C. Gagliardi, M. Khalil, and A. E. Foster, 'Streamlined Production of Genetically Modified T Cells with Activation, Transduction and Expansion in Closed-System G-Rex Bioreactors', *Cytotherapy*, vol. 21, no. 12, pp. 1246–1257, Dec. 2019, doi: 10.1016/j.jcyt.2019.10.006

[56] J. Ludwig and M. Hirschel, 'Methods and Process Optimization for Large-Scale CAR T Expansion Using the G-Rex Cell Culture Platform', in *Chimeric Antigen Receptor T Cells: Development and Production*, K. Swiech, K. C. R. Malmegrim, and V. Picanço-Castro, Eds., New York, NY: Springer US, 2020, pp. 165–177. doi: 10.1007/978-1-0716-0146-4_12

[57] H. K. Palani et al., 'Decentralized Manufacturing of Anti CD19 CAR-T Cells Using CliniMACS Prodigy®: Real-World Experience and Cost Analysis in India', *Bone Marrow Transplantation*, vol. 58, no. 2, Art. no. 2, Feb. 2023, doi: 10.1038/s41409-022-01866-5

[58] 'Technology', Cellares. Accessed: Mar. 14, 2024. [Online]. Available: https://www.cellares.com/technology/

[59] 'Cocoon® Platform | Cell Therapy Manufacturing | Lonza', Cocoon® Platform | Cell Therapy Manufacturing | Lonza. Accessed: Mar. 14, 2024. [Online]. Available: https://www.lonza.com/cell-and-gene/cocoon-platform

[60] A. Melocchi et al., 'Development of a Robotic Cluster for Automated and Scalable Cell Therapy Manufacturing'. bioRxiv, p. 2023.12.21.572854, Dec. 23, 2023. doi: 10.1101/2023.12.21.572854

[61] T. A. Smith, 'CAR-T Cell Expansion in a Xuri Cell Expansion System W25', in *Chimeric Antigen Receptor T Cells: Development and Production*, K. Swiech, K. C. R. Malmegrim, and V. Picanço-Castro, Eds., New York, NY: Springer US, 2020, pp. 151–163. doi: 10.1007/978-1-0716-0146-4_11

[62] L. Arnaudo, 'On CAR-Ts, Decentralized in-house Models, and the Hospital Exception. Routes for Sustainable Access to Innovative Therapies,' *Journal of Law and the Biosciences*, vol. 9, no. 2, p. lsac027, Sep. 2022, doi: 10.1093/jlb/lsac027

[63] N. Bäckel et al., 'Elaborating the potential of Artificial Intelligence in automated CAR-T cell manufacturing', *Frontiers in Molecular Medicine*, vol. 3, 2023, doi: 10.3389/fmmed.2023.1250508

[64] S. Hort et al., 'Toward Rapid, Widely Available Autologous CAR-T Cell Therapy – Artificial Intelligence and Automation Enabling the Smart Manufacturing Hospital', *Frontiers in Medicine*, vol. 9, 2022, doi: 10.3389/fmed.2022.913287

[65] A. Tay and N. Melosh, 'Mechanical Stimulation After Centrifuge-Free Nano-Electroporative Transfection Is Efficient and Maintains Long-Term T Cell Functionalities', *Small*, vol. 17, no. 38, p. 2103198, 2021, doi: 10.1002/smll.202103198

[66] E. Harris and J. J. Elmer, 'Optimization of Electroporation and Other Non-Viral Gene Delivery Strategies for T Cells', *Biotechnology Progress*, vol. 37, no. 1, p. e3066, 2021, doi: 10.1002/btpr.3066

6 Quality control of CAR-T products

Jerome Adriel Tjiptadi and Andy Kah Ping Tay

Introduction

Chimeric antigen receptor T cell (CAR-T) therapy has emerged as a promising avenue in cancer treatment, harnessing the power of the immune system to target and eliminate malignant cells. Thus far, a total of seven CAR-T products have been approved by the U.S. Food and Drug Administration (FDA), mainly for the treatment of a variety of hematologic malignancies and multiple myeloma [1]. With these successful outcomes, many novel CAR-T products are being explored and undergoing clinical trials globally [2]. With the rising number of CAR-T investigational new drug applications being submitted, there is a need for stringent quality control measures to guarantee product reliability and patient safety. On top of that, the complex nature of CAR-T manufacturing and the variability inherent in biological systems make quality control even more vital to ensure therapeutic efficacy while maintaining safety and consistency of CAR-T products.

In this chapter, we delve into the critical aspects of quality control in CAR-T product manufacturing. We aim to provide a comprehensive overview of the key product release testing and the emerging assays employed to assess the quality attributes of CAR-T cells.

General principles of CAR-T therapy quality control

The design, development, and testing of personalized "living" drugs such as CAR-T cells are challenging as they incorporate many different elements. The variability in CAR-T cell products can stem from the differences among donors such as age, sex, genetic background, and health status, including underlying medical conditions or prior treatments like chemotherapy and radiation therapy [3]. It is also significantly influenced by manufacturing conditions or even the quality of ancillary raw materials and reagents [4]. For instance, the proliferation, growth, and expansion of distinct phenotypic subset of T cells might be influenced by the differences in cell culture systems and conditions, which becomes a significant factor for designing CAR-T product quality control [5]. For example, Medvec et al. observed that fetal bovine serum (FBS), which is frequently used as a supplement in cell cultures, is able to improve CAR-T cell viability, proliferation, and expansion, but it lacks the consistency due to batch-to-batch variability [6, 7]. Hence, control of the manufacturing process and methodical testing in compliance with the current good manufacturing practices (cGMP) are required to warrant CAR-T products meet regulatory standards [8].

Following the completion of CAR-T cell manufacturing, validations through a wide range of quality control tests are crucial to ensure a consistent proof of quality and safety

DOI: 10.1201/9781032660752-6

Table 6.1 Guidelines related to CAR-T therapy quality control

Regulatory body	Guideline documents
FDA	• Consideration for the Development of Chimeric Antigen Receptor (CAR) T Cell Products – Guidance for Industry (01/2024) [11] • Potency Assurance for Cellular and Gene Therapy Products – Draft Guidance for Industry (12/2023) [16] • Chemistry, Manufacturing and Control (CMC) Information for Human Gene Therapy Investigational New Drug Applications (INDs) – Guidance for Industry (01/2020) [17] • Testing of Retroviral Vector-Based Human Gene Therapy Products for Replication Competent Retrovirus During Product Manufacture and Patient Follow-Up – Guidance for Industry (01/2020) [18]
EMA	• Guideline on quality, non-clinical and clinical aspects of medicinal products containing genetically modified cells (11/2020) [19] • Guideline on the quality, non-clinical and clinical aspects of gene therapy medicinal products (03/2018) [20] • Guideline on potency testing of cell-based immunotherapy medicinal products for the treatment of cancer (07/2016) [21]
ICH	• ICH Q2(R2) Validation of Analytical Procedures • ICH Q4B Annex 8 Pharmacopoeias – Sterility Test General Chapter • ICH Q4B Annex 14 Pharmacopoeias – Bacterial Endotoxins General Chapter • ICH Q5A Viral Safety Evaluation of Biotechnology Products Derived from Cell Lines of Human or Animal Origin • ICH Q6B Specifications: Test Procedures and Acceptance Criteria for Biotechnological/Biological Products

Abbreviations: FDA, Food and Drug Administration; EMA, European Medicines Agency; ICH, International Council for Harmonization of Technical Requirements for Pharmaceutical for Human Use.

performance that meets the predetermined release specifications. Generally, release testing should provide ample evidence that only the target cells are isolated, that they exhibit viability and functionality, and that they are free from contamination by other cells, microorganisms, or raw materials and reagents [9, 10]. The FDA has recently provided guidance and recommendation for the analytical assessment of CAR-T products [11]. The European Union (EU) through European Medicines Agency (EMA) has also made progress in putting down some framework to regulate advanced therapy medicinal products (ATMPs), in which CAR-T therapy is included. Several quality control guidelines from the FDA and EMA, along with documents from other regulatory bodies, are depicted in Table 6.1 [12]. However, they do not clearly establish a standardized procedure nor an authorized binding obligation, and there is no universally agreed-upon benchmark for CAR-T quality control. The quality metrics and attributes required for CAR-T products broadly include safety, purity, identity, and potency, summarized in Table 6.2 [13–15].

Safety

Safety stands as the cornerstone of CAR-T product release testing by prioritizing patient well-being and minimizing adverse events. Comprehensive safety assessments encompass a spectrum of evaluations, including tests to detect dangerous microbiological contaminants, such as endotoxins, mycoplasma, and other adventitious agents. Tests for the

Table 6.2 Critical quality attributes for CAR-T products

Category	Parameter	Assay	Advantages/limitations
Safety	Sterility	Compendial culture method	Standard, time-consuming
		Advanced culture method • BD BacTec™, BioMeriuex BacT/ ALERT3D® • Machine learning + long-read sequencing	Faster process, automated Identify low-abundance contaminants
		Non-culture method – gram staining	Faster, but less sensitive
	Mycoplasma	PCR	Rapid detection, but low sensitivity and labor intensive
	Endotoxin	Gel clot	Simple, long incubation time
		Turbidimetric	Faster, may be affected by turbidity
		Colorimetric	Easy to automate, subjective
		Chromogenic	Highly sensitive and specific, limited availability of chromogenic substrates
	Vector copy number	Endosafe Portable Testing System	Reproducible, accurate
		qPCR	Specific and sensitive
Purity	Process-related impurities, residual DNA	ELISA	Highly specific and sensitive, time-consuming and labor-intensive
		PCR	Highly specific and sensitive, require nucleic acid extraction
		Colorimetric assay	Simple, rapid, lacks specificity and precision
	Residual beads	Morphology assay – cell count	Simple, less accurate
		Flow cytometry	Accurate, reproducible
Identity	CAR expression, T cell phenotype	Flow cytometry	Phenotypic identification, accurate
		Imaging flow cytometry	Single-cell resolution
		Luciferase-based assay	Simple and versatile, lacks specificity
Potency	In vitro cytotoxicity	^{51}Cr assay	High reproducibility, short time required, but radioactive
		Bioluminescence-based assay	Non-radioactive, quick and robust, simple quantification, but potential inaccuracy
		Impedance-based assay	Label-free, real-time monitoring
		Flow cytometry	Highly sensitive, labor intensive
	Cytokine secretion profile	ELISA	Highly specific and sensitive, time-consuming and labor-intensive
		Flow cytometry	Highly sensitive, labor intensive
		ELISpot	Most sensitive, but time-consuming and complex
	VCN	qPCR	Highly sensitive and repeatable
		Digital PCR (dPCR)	Improved sensitivity, expensive
		Digital droplet PCR (ddPCR)	Low limit of detection, but potential obstruction by poor transduction efficiency

Abbreviations: PCR, polymerase chain reaction; LAL, limulus amebocyte lysate; DNA, deoxyribonucleic acid; ^{51}Cr, chromium; ELISA, enzyme-linked immunosorbent assay, ELISpot, enzyme-linked immunospot.

absence of replication competent retrovirus or lentivirus and vector copy number (VCN) should also be conducted when viral vectors are used, typically quantified using polymerase chain reaction (PCR) [10, 15]. Residual contamination during the manufacturing of engineered CAR-T cells can cause serious health implications, even death. For instance, a patient experienced relapse after CAR-T cell infusion due to the transduction of a single contaminating leukemic cell that expressed anti-CD19 CAR, which subsequently resulted in resistance to the CAR-T therapy [22].

Testing of sterility in aerobic and anaerobic bacteria, as well as fungi, can be done through both culture and non-culture methods, according to the United States Code of Federal Regulation (CFR) on "General Biological Product Standards" (21CFR610.12). In the culture-based method for sterility assessment, monitoring and observation for microbial growth is performed following the inoculation of the test samples into the culture media over a predefined period [9]. Recent technologies, such as BD BacTec™ and BioMeriuex BacT/ALERT 3D® systems, allow automated culture and microbial detection testing for a faster process of sterility testing. These systems utilize fluorescent or colorimetric carbon dioxide (CO_2) sensors for the continuous detection of microbial metabolic activity and will determine the presence of microorganisms based on the rate and amount of CO_2 increment in the system [23]. On the other hand, the non-culture-based method uses gram staining for faster but less-sensitive result compared to the culture-based testing [24, 25]. Among other quality control attributes, sterility test consumes the greatest amount of time, and hence, there is a need to develop rapid testing assays that are accurate and reliable [26]. Recently, Strutt et al. developed a machine learning-based microbial detection combined with long-read sequencing that is able to identify contaminants, even the low-abundance ones, within 24 hours [27]. More progress in this area would considerably expedite product release and enhance overall scalability.

One of the most common detection assays for mycoplasma is PCR, owing to its rapid detection compared to compendial culture method. However, this method might have some drawbacks, including low detection sensitivity and being labor-intensive. Sung et al. introduced a novel real-time PCR assay for culture medium that eliminates the necessity for DNA extraction, which demonstrates results of high specificity, sensitivity, and detection accuracy within 24 hours [28]. Another study by Dreolini et al. developed nucleic acid amplification techniques utilizing *Mycoplasma* genomic DNA instead of its live species to improve the limit of detection while reducing the duration required to produce results [29].

Bacterial endotoxin testing is conducted to identify substances released by lysed bacteria, such as gram-negative lipopolysaccharide, which could potentially induce harmful ramifications. The FDA guidelines state that the upper limit of the endotoxin level of the final product is 5 EU/kg. Limulus amebocyte lysate (LAL) test is the widely utilized method for endotoxin assessment, which encompasses four different release assays: one qualitative (i.e., gel clot) and three quantitative methods (i.e., turbidimetric, colorimetric, and chromogenic) [30]. An FDA-approved rapid testing called Endosafe Portable Testing System is proven to provide reproducible and accurate detection of endotoxin [31].

Testing of VCN is vital to assess the risk of insertional mutagenesis, where the viral vector integrates into critical genomic regions, potentially disrupting the normal cellular function or causing adverse effects [32]. Although theoretically there might be a positive correlation between the number of integrated transgene copies and the effectivity of CAR-T product [33], there is also an increased risk of oncogenic activity and genotoxicity with higher copy numbers, considering the randomness of vector integration into the host genome [34]. Thus, VCN should be controlled and tested to be in a safe but effective range of fewer than five vector copies per transduced cell as per FDA guidelines [35]. Besides, quantification of CAR transgene is critical to enable the exclusion of unnecessary transgenes (e.g., antibiotic resistance genes used for plasmid selection) from the vector, since it could cause unwanted consequences on the persistence and activity of the CAR-T product [11, 15]. Recently, the FDA has released an initiative to update the boxed warning on all approved CAR-T cell therapies with information regarding the significant risk of T cell malignancies [36]. Hence, VCN will also be an important factor that will influence cancer patients' decision for CAR-T treatment. The assays used for VCN quantification will be discussed further in the subsequent section as VCN can also be a surrogate measure to CAR-T product potency.

Purity

The manufacturing of CAR-T cell makes use of specific ancillary substances, such as antibodies, cytokines, sera, reagents for gene-editing, growth factors, culture and formulation medium, as well as cryopreservation reagents [4]. Purity assessment in CAR-T product release testing focuses on confirming the absence of unwanted residual cellular components or impurities related to the manufacturing process that could alter the therapeutic effect or trigger immune reactions [13, 15]. For instance, the presence of platelets, plasma, and residual anticoagulants may disrupt T cell response to activation reagents, while monocytes and granulocytes may impede the expansion and transduction of T cells in the process of CAR-T manufacturing [25, 37].

Other process-related impurities, namely bovine serum albumin, and benzonase, which is utilized for degradation and removal of DNA, as well as residual plasmid DNA and host cell DNA, need to be tested [38]. Various assays can be used to quantify these residual reagents, such as enzyme-linked immunosorbent assay (ELISA), quantitative PCR (qPCR), or colorimetric assays. Both ELISA and qPCR are highly specific and sensitive in their detection, but ELISA is time-consuming and labor-intensive, while qPCR requires nucleic acid extraction. Colorimetric assays (e.g., bicinchoninic acid assay) are superior in simplicity and rapid detection, but lack specificity and precision compared to the other two.

The manufacturing of many CAR-T products makes use of super-paramagnetic anti-CD3/CD28 antibody-coated beads (e.g., Dynal Dynabeads, human T-activator) as a tool for T cell activation and expansion. This activation step ensures that T cells are in an optimal state for CAR transduction and targeting of cancer cells [39, 40]. It is critical to ensure effective removal of beads (fewer than 100 beads per 3×10^6 cells) prior to infusion back to the patients due to the potential of polyclonal T cell activation [24], which may result in non-specific or excessive immune activation, and potentially lead to adverse effects if not carefully controlled [25]. Hollyman et al. adopted a morphology assay to count the number of residual Dynabeads [24], whereas Schalkwyk et al. used a novel flow cytometric assay that is able to produce accurate and reproducible quantification of residual Dynabeads [41].

Identity

Identity testing verifies the authenticity and consistency of CAR-T product throughout the manufacturing process. It ensures that the final product matches the intended specifications and retains its characteristic features, including physicochemical properties (i.e., visual appearance, pH, osmolality), presence of CAR transgene (i.e., CAR expression), and cellular composition (i.e., specific T cell markers) [11, 42].

Different phases of T cell development, such as differentiation, activation, and memory formation, can be indicated by the distinction in its surface markers. Past research and clinical studies found that the profile of CAR-T surface markers corresponds to its therapeutic efficacy [43]. In general, CAR-T cells exhibiting a memory-like phenotype, characterized by CCR7, CD62L, CD45RA, CD45RO, and CD95 surface markers, demonstrate heightened potency [44, 45]. Conversely, the CAR-T cells with exhaustion-like phenotype, expressed by inhibitory receptors like PD-1, TIM-3, LAG-3, CTLA4, and TIGIT, show reduced effectiveness [25, 46, 47].

The memory T cells can also be divided into different subpopulations based on their phenotypic differences and combination of markers expressed on their surface: naïve T cells (T_N), central memory T cells (T_{CM}), effector memory T cells (T_{EM}), terminally differentiated effector memory T cells (T_{EMRA}), stem cell-like memory T cells (T_{SCM}), and resident memory T cells (T_{RM}). Effector T cell subsets, T_{EM} and T_{EMRA}, have been known to display superior in vitro cytotoxicity and cytokine-releasing potential to kill tumors. However, their low level of self-renewal capacity compared to the less-differentiated T_{CM} subset hinders them from having effective proliferative, persistence, and in vivo functionality [39, 48–50]. Meanwhile, T_{SCM} self-renewal ability and differentiation capacity allow effective in vivo expansion and persistence [51]. Therefore, controlling and setting up metrics of the proportion of different favorable T cell subtypes is vital for the potency and persistence of CAR-T products, and will benefit the development of future commercial products [42].

Quantification of CAR protein expression as well as the different subpopulations of CAR-T cells has been mainly evaluated by flow cytometry. Flow cytometry can characterize immunophenotypic T cell markers such as CD3, CD4, CD8, CD15, CD19, and CD56 [52–54]. Flow cytometry is preferred here as opposed to qPCR, since qPCR may overestimate the amount of functional CAR-T cells and provide no information regarding CAR-T cells' phenotype [55]. The majority of flow cytometry analyses to determine CAR expression make use of labeled anti-CAR antibodies and measuring its fluorescence intensity [52–54]. Various commercially available staining agents, such as Protein L, CD19-Fc, anti-IgG F(ab')2 antibodies, and anti-CD19 FMC63 idiotype antibodies, can target surface CAR molecules, each with their own benefits and disadvantages in terms of specificity, stability, compatibility, and cost [55, 56]. Besides, reporter gene expression can serve as a positive selection marker for CAR+ cells [57].

Another emerging method for CAR detection that is presently still in development phase is imaging flow cytometry, which combines the capability of conventional flow cytometry and digital fluorescence microscopy. Incorporating digital microscopy into the arsenal for CAR detection allows for the observation of distinct morphological and phenotypic characteristics of CAR-T at single-cell resolution, including the spatial arrangement of surface CAR protein or intracellular reporter gene. Consequently, it becomes a promising methodology to thoroughly verify the identity and consistency of CAR-T products [15, 55].

The detection of CAR expression and binding ability can also be assessed with high specificity and sensitivity using a luciferase-based assay utilizing the Topanga reagent, developed by Gopalakrishnan et al. This assay is simple and versatile as it only involves incubation of CAR-T cells with the Topanga reagent and detection by luminescence. However, there is a need to develop the Topanga reagent further so it can be used specifically for different CAR constructs (e.g., CD19, CD138) [58].

Potency

Potency assessment aims to quantify functional activity of CAR-T cells and predict their therapeutic efficacy in vivo. This involves evaluating the in vitro cytotoxicity against target cancer cells, cytokine (i.e., IFN-γ) secretion profiles, and proliferative capacity of the CAR-T product [4]. Transduction efficiency, CAR expression, VCN, and vector integration are also normally used as representative measures of CAR-T product potency [42]. Viability testing with a release criterion of at least 70% is also important for CAR-T efficacy, although some research indicated that there are no differences in therapeutic outcome for CAR-T product with 70%–80% viability and more than 80% viability [59, 60]. The FDA recommended the use of orthogonal methodology of potency testing (e.g., cell killing assay, cytokine release assay, and transduction efficiency assay), which quantitatively measures the true value of certain product attributes using diverse physical principles, to reduce the risk of measurement bias or interference [11, 61, 62].

There are several cytotoxic assays most frequently used to provide insights to the cytolytic activity of CAR-T cells against tumor targets, including chromium (^{51}Cr) release assay, luciferase-mediated bioluminescence imaging, impedance-based assay, and flow cytometry [63] (Figure 6.1). ^{51}Cr assay has been the gold standard for cytotoxicity measurement due to its high reproducibility and relatively shorter time required. This assay involves labeling target cells with radioactive chromium and incubating them together with effector cells at various effector-to-target (E:T) ratios, then measuring the cytotoxic activity from the release of chromium into the supernatant upon lysis by the effector cells. However, this assay is limited to a single time point measurement, and the radioactivity of ^{51}Cr is potentially harmful to the assay user [63]. Alternatively, bioluminescence-based cytotoxic assay offers a non-radioactive option to perform a quicker and more robust measurement of CAR-T cytotoxicity [64, 65]. A bioluminescence assay quantifies cytotoxicity by measuring the reduction in photons emitted from target cells expressing a luciferase reporter gene upon lysis. This method can be a reliable proxy for CAR-T cell efficacy with simple handling and quantification, but may not be feasible for every cell type as well as may have the potential of exaggeration in its cytotoxicity value [66]. Another methodology, an impedance-based assay, measures the changes in electrical impedance caused by cell adhesion, enabling a label-free and dynamic real-time monitoring of cytolytic activity over a certain duration. This assay is advantageous also owing to its enhanced sensitivity and capability to operate at low E:T ratios [67, 68].

Flow cytometry is able to quantify the cytotoxicity of cell subpopulation within a heterogeneous cell mixture with great sensitivity. It works by distinguishing between the effector and target cell based on their size and granularity, as well as stain of different antibodies. However, flow cytometry–based assays are labor-intensive due to the necessity to acquire and analyze individual data [63]. Recently, there has been progress in the development of flow cytometry to produce high-throughput and multiparametric results for more comprehensive analysis of cytotoxicity and even other parameters like transduction efficiency and activity status [69]. Besides the aforementioned four common assays, many researchers are developing unique novel methodologies for cytotoxicity measurement, for instance, using droplet microfluidics [70] or 3D hanging spheroid plates [71]. However, these assays require further validations and fulfill requirements of accuracy, precision, reproducibility, specificity, and robustness according to the International Council for Harmonization of Technical Requirements for Pharmaceutical Human Use (ICH) guidelines.

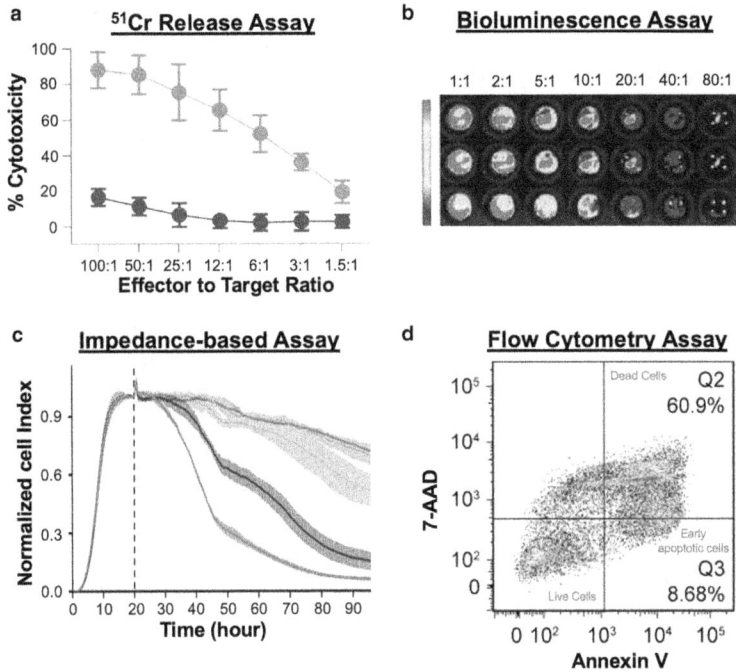

Figure 6.1 Comparison between readouts of various cytotoxicity assays to evaluate the potency of CAR-T therapy. (a) ^{51}Cr release assay. The release of chromium from labeled target cells cocultured with control cells (blue curve) or effector cells (red curve) is assessed with respect to a maximum and a spontaneous chromium release control. (b) Bioluminescence assay. The cytotoxicity mediated by the effector cells at different E:T ratios is indicated by the quantity of viable target cells expressing luciferase, which is correlated to the bioluminescence intensity. (c) Impedance-based assay. A decline in normalized cell index, indicating the detachment of adherent target cells, is observed subsequent to the addition of control cells (green curve) or effector cells (blue curve) at a particular E:T ratio and at a specific duration following the seeding of target cells (dotted line). The orange and red curves represent the untreated target cells and the 100% lysis control group, respectively. (c) Flow cytometry assay. The plot describes staining of target cells with 7-amino-actinomycin D (7-AAD) and Annexin V that is undergoing apoptosis and cell death, mediated by the effector cells. Figures adapted from Kiesgen et al. (2021) [63] with permission.

Assessment cytokine secretion profiles, particularly interleukin-2 (IL-2), interleukin-15 (IL-15) and interferon-gamma (IFN-γ), offer valuable information about the potency and activation status of CAR-T cells [43]. Furthermore, CAR-T therapy has a risk of systemic toxicity such as cytokine release syndrome (CRS), which may cause serious adverse effects. Thus, it is vital to understand the cytokine release characteristic for the safety of the patient [72]. The typical assays for cytokine release are ELISA, intracellular cytokine staining combined with flow cytometry, and enzyme-linked immunospot (ELISpot) assays. ELISpot offers the most sensitive detection compared to the other two, but it is also the most time-consuming and complex [73].

Besides being a consideration for safety, VCN analysis can be a measure of CAR-T product potency as it influences the level of CAR protein expression and transduction efficiency. qPCR is the most common assay for the highly sensitive and repeatable determination of CAR VCN [74]. Digital PCR (dPCR) offers improved sensitivity compared to qPCR due to single-cell VCN measurement. However, dPCR equipment and consumables are more expensive and uncommon [55]. Meanwhile, digital droplet PCR (ddPCR) allows even better sensitivity enabling the detection of one anti-CD19 CAR-T cell in a background of 10,000 cells, but it might be obstructed by poor CAR transduction efficacy [75, 76].

Conclusion

The complex nature of CAR-T cell manufacturing necessitates rigorous and extensive quality control in order for the final product to be approved and released for infusion back to patients. Numerous factors define the safety and efficacy of CAR-T products, including CAR design, vector delivery method, source of T cell, manipulation of the process, biological activities, and incorporation of ancillary novel components (e.g., gene editing, immunomodulatory elements). Besides, ancillary materials for CAR-T are vital to be controlled for quality. One example is viral vectors. There is a possibility for viral capsids to only be partially filled or even not filled with the genomic materials during their assembly, which may result in both positive and negative implications to the safety and efficacy of the product [77, 78]. Regardless of their ramification, the identification and quantification (i.e., viral titer, multiplicity of infection [MOI]) of viral capsids is critical to ensuring product safety and consistency. Various assays varying in their sensitivity, ease of use, and specificity can be employed for viral quantification, including plaque assay, 50% tissue culture infectious dose ($TCID_{50}$) assay qPCR, differential UV absorbance measurement, and electron microscopy [79]. Overall, the quality control of CAR-T product is not only required at the final stage before being released, but whole-process quality controls, including control of starting materials, in-process control and testing, and stability testing, are essential to ensuring good quality and efficacy of the product.

Moreover, cellular and molecular quality control attributes (i.e., safety, purity, identity, and potency) require time-consuming and laborious processes. There is also a deficit of suitable benchmark, reference materials, and performance controls that warrant repeatability and integration of testing. Therefore, as CAR-T therapy continues to be developed for wider clinical applications, a clear and comprehensive guideline for high-quality and effective CAR-T production is required. With novel advances in CAR-T technologies that result in more potent treatment, as well as developments in assays for the release testing of the products, there will be continual adjustment and test validations in the future to fine-tune the parameters and requirements for robust quality control.

Through this exploration of quality control strategies, assays, and metrics, we aim to provide valuable insights and practical guidance to researchers, clinicians, and manufacturers involved in the development and commercialization of CAR-T therapies. By fostering a deeper understanding of quality control principles and methodologies, we endeavor to advance the field of CAR-T therapy and contribute to the ongoing pursuit of safe, effective, and standardized cellular immunotherapies.

References

1. Cappell, K.M. and J.N. Kochenderfer, Long-term outcomes following CAR T cell therapy: what we know so far. *Nat Rev Clin Oncol*, 2023. **20**(6): p. 359–371.
2. Wang, V., et al., Systematic review on CAR-T cell clinical trials Up to 2022: academic center input. *Cancers (Basel)*, 2023. **15**(4).

3. Fesnak, A.D., The challenge of variability in chimeric antigen receptor T cell manufacturing. *Regen Eng Transl Med*, 2020. **6**(3): p. 322–329.
4. Wang, X. and I. Riviere, Clinical manufacturing of CAR T cells: foundation of a promising therapy. *Mol Ther Oncolytics*, 2016. **3**: p. 16015.
5. Guedan, S., H. Calderon, A.D. Posey, Jr., and M.V. Maus, Engineering and design of chimeric antigen receptors. *Mol Ther Methods Clin Dev*, 2019. **12**: p. 145–156.
6. Medvec, A.R., et al., Improved expansion and in vivo function of patient T cells by a serum-free medium. *Mol Ther Methods Clin Dev*, 2018. **8**: p. 65–74.
7. Sudarsanam, H., R. Buhmann, and R. Henschler, Influence of culture conditions on ex vivo expansion of T lymphocytes and their function for therapy: current insights and open questions. *Front Bioeng Biotechnol*, 2022. **10**: p. 886637.
8. Gee, A.P., GMP CAR-T cell production. *Best Pract Res Clin Haematol*, 2018. **31**(2): p. 126–134.
9. Gee, A.P., Manufacturing genetically modified T cells for clinical trials. *Cancer Gene Ther*, 2015. **22**(2): p. 67–71.
10. Reddy, O.L., D.F. Stroncek, and S.R. Panch, Improving CAR T cell therapy by optimizing critical quality attributes. *Semin Hematol*, 2020. **57**(2): p. 33–38.
11. U.S. Department of Health and Human Services. Considerations for development of chimeric antigen receptor (CAR) T cell products - guidance for industry. 2024 April 10, 2024]; Available from: https://www.fda.gov/media/156896/download
12. Sharpe, M. and N. Mount, Genetically modified T cells in cancer therapy: opportunities and challenges. *Dis Model Mech*, 2015. **8**(4): p. 337–350.
13. Levine, B.L., Performance-enhancing drugs: design and production of redirected chimeric antigen receptor (CAR) T cells. *Cancer Gene Ther*, 2015. **22**(2): p. 79–84.
14. Li, Y., Y. Huo, L. Yu, and J. Wang, Quality control and nonclinical research on car-t cell products: general principles and key issues. *Engineering*, 2019. **5**(1): p. 122–131.
15. Ramesh, P., et al., Chimeric antigen receptor T-cells: Properties, production, and quality control. *Int J Lab Hematol*, 2023. **45**(4): p. 425–435.
16. U.S. Department of Health and Human Services. Potency assurance for cellular and gene therapy products - draft guidance for industry. 2023 May 25, 2024]; Available from: https://www.fda.gov/media/175132/download
17. U.S. Department of Health and Human Services. Chemistry, manufacturing, and control (CMC) information for human gene therapy investigational new drug applications (INDs) - guidance for industry. 2020 May 25, 2024]; Available from: https://www.fda.gov/media/113760/download
18. U.S. Department of Health and Human Services. Testing of retroviral vector-based human gene therapy products for replication competent retrovirus during product manufacture and patient follow-up - guidance for industry. 2020 May 25, 2024]; Available from: https://www.fda.gov/media/113790/download
19. European Medicines Agency. EMA/CAT/GTWP/671639/2008 - Guideline on quality, non-clinical and clinical aspects of medicinal products containing genetically modified cells. 2020 May 26, 2024]; Available from: https://www.ema.europa.eu/en/documents/scientific-guideline/guideline-quality-non-clinical-and-clinical-aspects-medicinal-products-containing-genetically-modified-cells-revision-1_en.pdf
20. European Medicines Agency. EMA/CAT/80183/2014 - Guideline on the quality, non-clinical and clinical aspects of gene therapy medicinal products. 2018 May 26, 2024]; Available from: https://www.ema.europa.eu/en/documents/scientific-guideline/guideline-quality-non-clinical-and-clinical-aspects-gene-therapy-medicinal-products_en.pdf
21. European Medicines Agency. EMA/CHMP/BWP/271475/2006 - Guideline on potency testing of cell based immunotherapy medicinal products for the treatment of cancer. 2016 May 26, 2024]; Available from: https://www.ema.europa.eu/en/documents/scientific-guideline/guideline-potency-testing-cell-based-immunotherapy-medicinal-products-treatment-cancer-revision-1_en.pdf
22. Ruella, M., et al., Induction of resistance to chimeric antigen receptor T cell therapy by transduction of a single leukemic B cell. *Nat Med*, 2018. **24**(10): p. 1499–1503.
23. England, M.R., et al., Comprehensive evaluation of compendial USP<71>, BacT/Alert Dual-T, and Bactec FX for detection of product sterility testing contaminants. *J Clin Microbiol*, 2019. **57**(2): p. 10.1128/jcm.01548-18.

24. Hollyman, D., et al., Manufacturing validation of biologically functional T cells targeted to CD19 antigen for autologous adoptive cell therapy. *J Immunother*, 2009. **32**(2): p. 169–180.

25. Roddie, C., et al., Manufacturing chimeric antigen receptor T cells: issues and challenges. *Cytotherapy*, 2019. **21**(3): p. 327–340.

26. Abou-El-Enein, M., et al., Scalable Manufacturing of CAR T cells for Cancer Immunotherapy. *Blood Cancer Discov*, 2021. **2**(5): p. 408–422.

27. Strutt, J.P.B., et al., Machine learning-based detection of adventitious microbes in T-cell therapy cultures using long-read sequencing. *Microbiology Spectrum*, 2023. **11**(5): p. e01350–e01323.

28. Sung, J. and J.R. Hawkins, A highly sensitive internally-controlled real-time PCR assay for mycoplasma detection in cell cultures. *Biologicals*, 2020. **64**: p. 58–72.

29. Dreolini, L., et al., A rapid and sensitive nucleic acid amplification technique for mycoplasma screening of cell therapy products. *Mol Ther Methods Clin Dev*, 2020. **17**: p. 393–399.

30. Picanco-Castro, V., et al., Associacao brasileira de hematologia, hemoterapia e terapia celular consensus on genetically modified cells. VIII: CAR-T cells: preclinical development - Safety and efficacy evaluation. *Hematol Transfus Cell Ther*, 2021. **43 Suppl 2**(Suppl 2): p. S54–S63.

31. Barry, N., Z. Velickovic, and J. Rasko, Endotoxin quality control testing for CAR T-cell manufacturing: validation considerations for Endosafe portable testing system. *Cytotherapy*, 2020. **22**(5, Supplement): p. S140.

32. Vannucci, L., et al., Viral vectors: a look back and ahead on gene transfer technology. *New Microbiol*, 2013. **36**(1): p. 1–22.

33. Maude, S.L., et al., Chimeric antigen receptor T cells for sustained remissions in leukemia. *N Engl J Med*, 2014. **371**(16): p. 1507–1517.

34. Chang, A.H. and M. Sadelain, The genetic engineering of hematopoietic stem cells: the rise of lentiviral vectors, the conundrum of the ltr, and the promise of lineage-restricted vectors. *Mol Ther*, 2007. **15**(3): p. 445–456.

35. U.S. Department of Health and Human Services. Long term follow-up after administration of human gene therapy products - guidance for industry. 2020 April 10, 2024]; Available from: https://www.fda.gov/media/113768/download

36. U.S. Department of Health and Human Services. FDA requires boxed warning for T cell malignancies following treatment with BCMA-directed or CD19-directed autologous chimeric antigen receptor (CAR) T cell immunotherapies. 2024 April 18, 2024 [cited May 8, 2024; Available from: https://www.fda.gov/vaccines-blood-biologics/safety-availability-biologics/fda-requires-boxed-warning-t-cell-malignancies-following-treatment-bcma-directed-or-cd19-directed

37. Stroncek, D.F., et al., Myeloid cells in peripheral blood mononuclear cell concentrates inhibit the expansion of chimeric antigen receptor T cells. *Cytotherapy*, 2016. **18**(7): p. 893–901.

38. Levine, B.L., J. Miskin, K. Wonnacott, and C. Keir, Global manufacturing of CAR T cell therapy. *Mol Ther Methods Clin Dev*, 2017. **4**: p. 92–101.

39. Barrett, D.M., et al., Relation of clinical culture method to T-cell memory status and efficacy in xenograft models of adoptive immunotherapy. *Cytotherapy*, 2014. **16**(5): p. 619–630.

40. Green, L., Development of an anti-CD2/CD3/CD28 bead-based T-cell proliferation assay. *Biosci Horiz*, 2014. **7**.

41. van Schalkwyk, M.C.I., et al., Development and validation of a good manufacturing process for IL-4-driven expansion of chimeric cytokine receptor-expressing CAR T-cells. *Cell*, 2021. **10**(7). Article Number: 1797.

42. Ayala Ceja, M., et al., CAR-T cell manufacturing: Major process parameters and next-generation strategies. *J Exp Med*, 2024. **221**(2). Article Number: e20230903.

43. Si, X., L. Xiao, C.E. Brown, and D. Wang, Preclinical evaluation of CAR T cell function: in vitro and in vivo models. *Int J Mol Sci*, 2022. **23**(6). Article Number: 3154.

44. Alizadeh, D., et al., IL15 enhances CAR-T Cell antitumor activity by reducing mTORC1 activity and preserving their stem cell memory phenotype. *Cancer Immunol Res*, 2019. **7**(5): p. 759–772.

45. Fraietta, J.A., et al., Determinants of response and resistance to CD19 chimeric antigen receptor (CAR) T cell therapy of chronic lymphocytic leukemia. *Nat Med*, 2018. **24**(5): p. 563–571.

46. Long, A.H., et al., 4-1BB costimulation ameliorates T cell exhaustion induced by tonic signaling of chimeric antigen receptors. *Nat Med*, 2015. **21**(6): p. 581–590.

47. Cherkassky, L., et al., Human CAR T cells with cell-intrinsic PD-1 checkpoint blockade resist tumor-mediated inhibition. *J Clin Invest*, 2016. **126**(8): p. 3130–3144.
48. McLellan, A.D. and S.M. Ali Hosseini Rad, Chimeric antigen receptor T cell persistence and memory cell formation. *Immunol Cell Biol*, 2019. **97**(7): p. 664–674.
49. Feucht, J., et al., Calibration of CAR activation potential directs alternative T cell fates and therapeutic potency. *Nat Med*, 2019. **25**(1): p. 82–88.
50. Xu, Y., et al., Closely related T-memory stem cells correlate with in vivo expansion of CAR. CD19-T cells and are preserved by IL-7 and IL-15. *Blood*, 2014. **123**(24): p. 3750–3759.
51. Tantalo, D.G., et al., Understanding T cell phenotype for the design of effective chimeric antigen receptor T cell therapies. *J Immunother Cancer*, 2021. **9**(5).
52. Peinelt, A., et al., Monitoring of circulating CAR T cells: Validation of a flow cytometric assay, cellular kinetics, and phenotype analysis following tisagenlecleucel. *Front Immunol*, 2022. **13**: p. 830773.
53. Sarikonda, G., et al., Best practices for the development, analytical validation and clinical implementation of flow cytometric methods for chimeric antigen receptor T cell analyses. *Cytometry B Clin Cytom*, 2021. **100**(1): p. 79–91.
54. Blache, U., et al., Advanced flow cytometry assays for immune monitoring of CAR-T cell applications. *Front Immunol*, 2021. **12**: p. 658314.
55. Hu, Y. and J. Huang, The chimeric antigen receptor detection toolkit. *Front Immunol*, 2020. **11**: p. 1770.
56. Schanda, N., et al., Sensitivity and specificity of CD19.CAR-T cell detection by flow cytometry and PCR. *Cell*, 2021. **10**(11).
57. Rodriguez-Marquez, P., et al., CAR density influences antitumoral efficacy of BCMA CAR T cells and correlates with clinical outcome. *Science Advances*, 2022. **8**(39): p. eabo0514.
58. Gopalakrishnan, R., et al., A novel luciferase-based assay for the detection of Chimeric Antigen Receptors. *Sci Rep*, 2019. **9**(1): p. 1957.
59. Chong, E.A., et al., Impact of CAR T-cell product viability on B-cell lymphoid malignancy outcomes. *Cytotherapy*, 2019. **21**(5, Supplement): p. S19.
60. Chong, E.A., et al., CAR T cell viability release testing and clinical outcomes: is there a lower limit? *Blood*, 2019. **134**(21): p. 1873–1875.
61. Simon, C.G., Jr., et al., Orthogonal and complementary measurements of properties of drug products containing nanomaterials. *J Control Release*, 2023. **354**: p. 120–127.
62. Simon, C.G., Jr., et al., Mechanism of action, potency and efficacy: considerations for cell therapies. *J Transl Med*, 2024. **22**(1): p. 416.
63. Kiesgen, S., et al., Comparative analysis of assays to measure CAR T-cell-mediated cytotoxicity. *Nat Protoc*, 2021. **16**(3): p. 1331–1342.
64. Karimi, M.A., et al., Measuring cytotoxicity by bioluminescence imaging outperforms the standard chromium-51 release assay. *PLoS One*, 2014. **9**(2): p. e89357.
65. Matta, H., et al., Development and characterization of a novel luciferase based cytotoxicity assay. *Sci Rep*, 2018. **8**(1): p. 199.
66. Brakel, B.A., et al., In vitro evaluation of CAR-T cells in patient-derived glioblastoma models. *STAR Protoc*, 2021. **2**(4): p. 100920.
67. Peper, J.K., et al., An impedance-based cytotoxicity assay for real-time and label-free assessment of T-cell-mediated killing of adherent cells. *J Immunol Methods*, 2014. **405**: p. 192–198.
68. Cao, M., R.D. Carlson, R.E. Staudt, and A.E. Snook, In vitro assays to evaluate CAR-T cell cytotoxicity. *Methods Cell Biol*, 2024. **183**: p. 303–315.
69. Martinez, E.M., et al., High-throughput flow cytometric method for the simultaneous measurement of CAR-T cell characterization and cytotoxicity against solid tumor cell lines. *SLAS Discov*, 2018. **23**(7): p. 603–612.
70. Wong, K.U., et al., Assessment of chimeric antigen receptor T cytotoxicity by droplet microfluidics in vitro. *Antib Ther*, 2022. **5**(2): p. 85–99.
71. Chen, Z., et al., 3D hanging spheroid plate for high-throughput CAR T cell cytotoxicity assay. *J Nanobiotechnology*, 2022. **20**(1): p. 30.
72. Brudno, J.N. and J.N. Kochenderfer, Toxicities of chimeric antigen receptor T cells: recognition and management. *Blood*, 2016. **127**(26): p. 3321–3330.
73. Wagner, D.L., et al., Immunogenicity of CAR T cells in cancer therapy. *Nat Rev Clin Oncol*, 2021. **18**(6): p. 379–393.

74. Kunz, A., et al., Optimized assessment of qPCR-based vector copy numbers as a safety parameter for GMP-grade CAR T cells and monitoring of frequency in patients. *Mol Ther Methods Clin Dev*, 2020. **17**: p. 448–454.

75. Mika, T., et al., Digital-droplet PCR for quantification of CD19-directed CAR T-cells. *Front Mol Biosci*, 2020. **7**: p. 84.

76. Haderbache, R., et al., Droplet digital PCR allows vector copy number assessment and monitoring of experimental CAR T cells in murine xenograft models or approved CD19 CAR T cell-treated patients. *J Transl Med*, 2021. **19**(1): p. 265.

77. Wright, J.F., AAV empty capsids: for better or for worse? *Mol Ther*, 2014. **22**(1): p. 1–2.

78. Wright, J.F., Manufacturing and characterizing AAV-based vectors for use in clinical studies. *Gene Ther*, 2008. **15**(11): p. 840–848.

79. Lothert, K., E. Bagrin, and M.W. Wolff, Evaluating novel quantification methods for infectious baculoviruses. *Viruses*, 2023. **15**(4). Article Number: 998.

7 Advances *in vivo* CAR-T manufacturing

Mohamed Bilal S/O Saibudeen, Kenneth Jun Lim Goh, and Andy Kah Ping Tay

Introduction

Chimeric antigen receptor (CAR)-T cell therapy is a form of immunotherapy that has shown remarkable clinical responses against B-cell malignancies and multiple myeloma [1]. To date, the US Food and Drug Administration (FDA) approved six CAR-T cell therapies: Kymriah, Yescarta, Tecartus, Breyanzi, Abecma, and Carvykti [2]. Among these, the first four CAR-T cell products target Cluster of Differentiation (CD) 19 and the latest two target B-cell maturation antigens (BCMA) [2]. However, all market-approved therapies use autologous CAR-T cells, and their individualized manufacturing process is highly complex and can cost as much as US$500k/dose and take up to five weeks to manufacture, which is untimely for cancer types with rapid progression [3, 4, 8]. Additionally, only specialized institutions have the resources and technical knowledge to manufacture [7, 8]. Current CAR-T cell therapy also faces several other barriers such as associated toxicity and poor results against solid tumors [6, 9].

To simplify the manufacturing process for overcoming inherent limitations of the current treatment, one promising method involves a shift from the traditional ex vivo methodology to an in vivo approach where CAR genes are delivered to T cells in the patient's body followed by in vivo expansion [7]. This approach circumvents the time-consuming and complex ex vivo manufacturing process (Figure 7.1a). In this chapter, we will discuss in greater detail emerging technologies for in vivo transfection and expansion.

In vivo transfection

To circumvent the limitations associated with ex vivo manufacturing of autologous CAR-T cells, researchers are developing methods for in vivo generation of CAR-T cells [2, 9]. Ex vivo T cell manufacturing requires extensive rounds of cell expansion, causing persistent activation and stimulation, which is associated with exhaustion and differentiation into less-effective cellular subtypes such as terminal effector T cells [1, 5, 21]. In contrast, it has been shown in preclinical models that in situ programming of T cells reduces cell exhaustion and leads to a greater proportion of naïve T cells, which can mature and differentiate into long-lived memory T cells, hence improving persistence and contributing to prolonged remission [1, 7, 21]. To perform in vivo transfection, viruses and nanoparticles are used to deliver genes into T cells.

Viral vectors are predominantly used for ex vivo transduction in clinically approved CAR-T cell therapies due to their high transduction efficacy [4, 10] (Figure 7.1b). Commonly used viral vectors include lentiviruses and gamma-retroviral systems.

DOI: 10.1201/9781032660752-7

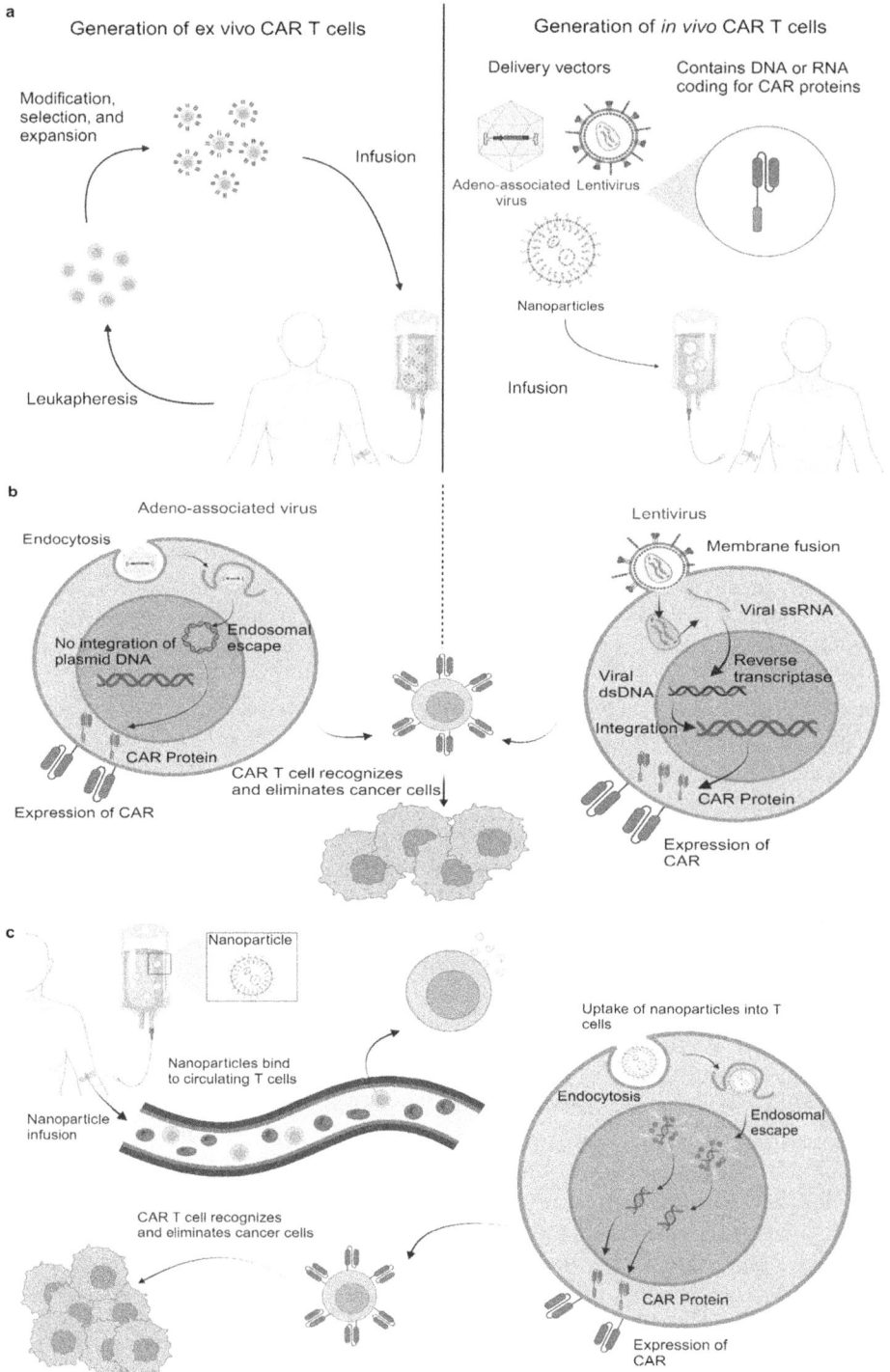

Figure 7.1 The use of nanoparticles and viral vectors for in vivo CAR-T transduction. (a) In vivo vs ex vivo CAR-T manufacturing. Inspired by [10]. (b) Schematic showing in vivo transduction using viral vectors. (c) Schematic showing in vivo transfection using nanoparticles. Inspired by [17].

The limited number of GMP production facilities for viral vectors has led to costly treatments, long manufacturing lead times, and a global shortage of CAR-T products [4, 25]. An alternative non-viral delivery method involves nanoparticles between 100 and 1,000 nm in diameter [22] (Figure 7.1c). Nanoparticles are highly versatile because they can be synthesized with differing characteristics, including lipids, polymers, proteins, and inorganic materials [4, 10]. Moreover, through design specifications, nanocarriers for in vivo gene delivery can achieve their desired physicochemical properties to ensure the high efficacy of the treatment [10]. In vivo CAR-T transfection may offer advantages such as better safety, lower product price, availability as off-the-shelf prescription, and no preconditioning of patients [4, 7].

Advantages of in vivo CAR-T transfection

In vivo CAR-T transfection could result in reduction of systemic toxicity. A problem with CAR-T therapy is systemic toxicity such as cytokine release syndrome (CRS) and immune-effector cell-associated neurotoxicity syndrome (ICANS), which have prevented broader implementation of the therapy as some very ill patients are unlikely to withstand these side effects. CRS is a response triggered after CAR-T cells are activated and have over-produced inflammatory cytokines, including IL-6, IL-10, IL-2, and TNFα [9]. The principal cytokine that causes CRS is IL-6 and is the main target during treatment [2, 13]. CRS commonly causes fever, muscle aches, cold, and hypotension but can lead to loss of function in organs in more serious cases [2, 9, 13]. The next most common systemic toxicity associated with CAR-T therapy is ICANS, which occurs concurrently or after recovery from CRS. The effects of ICANS manifest as toxic encephalopathy causing neurological complications, such as aphasia, confusion, and difficulty in finding words [9, 13].

Effective in vivo transduction has been achieved with adeno-associated viruses (AAV) and lentiviral vectors [5, 9, 24]. Lai et al. successfully engineered a mutated Sindbis-based lentiviral vector to express bispecific binders encoding second-generation CD19-specific CAR, reducing off-target transduction, and redirecting the vector to T cells [24]. The lentiviral vector effectively performed in vivo transduction in murine models, and even with limited CAR-T cells, a reduction was observed in tumor cells [24]. They used this as proof that in vivo CAR-T cells show greater performance and self-renewal capacity over ex vivo CAR-T cells [24]. Wu et al. created a model using an AAV vector encoding a third-generation CAR that achieved cancer remission in murine models with a single infusion [5]. However, in this study their main objective was to propose the possibility of AAV vectors without considering the specificity and potential off-target effects of viral transduction. Nevertheless, viral vectors have limitations, including (i) the adverse effect of off-target transduction, (ii) the possibility of insertional mutagenesis, and (iii) the restricted cargo capacity [4]. To overcome these disadvantages, researchers are looking at alternative material such as nanoparticles to perform in vivo CAR-T transfection.

Nanoparticles are chemically created, making them highly adaptable to meet therapeutic requirements by varying their compositions accordingly [4]. Moreover, nanocarriers can draw inspiration from viral vectors, replicating their physical structure to achieve the same properties at a lower cost, due to their quality being easier to ensure [4]. Furthermore, nanocarriers can contain not only CAR genes but also DNA editing tools or other compounds, enabling a mixture of elements that complement one another to be delivered together, enhancing the overall therapy [7, 9]. Moreover, nanomaterials are chemically stable and are easy to store for prolonged periods for off-the-shelf treatments [4, 7]. In addition, with ongoing advancements in nanoparticle manufacturing technology, the process of creating them could be fully automated, thus reducing product costs [15]. Stephan et al. developed a DNA nanocarrier consisting of two plasmids containing either an anti-CD19 CAR or hyperactive piggyBac transposase enzyme (IBP7) plasmids encapsulated within a core composed of biodegradable poly (β-amino ester)-based nanoparticles, while the exterior was composed of a mixture of polyglutamic acid and anti-CD3 antibody [7]. These nanoparticles managed to achieve remission in murine models with no signs of system toxicities [7]. Another experiment by Buchholz et al. used a lentiviral vector that resulted in systemic toxicities similar to those observed during clinical trials [12]. However, a notable difference is that CRS happened slower than those in clinical trials which the researchers hypothesized to be attributed to the number of CAR-T cells increasing slowly by expansion as opposed to an infusion of a large number of ex vivo manufactured CAR-T cells [12]. This suggests that CRS could be better managed when cytokine levels rise steadily rather than a sudden spike.

In vivo CAR-T transfection could also remove the need for patient preconditioning. Patients undergo lymphodepleting chemotherapy a week before infusion as required for current CAR-T therapy [1, 16]. Research has shown that this preconditioning remarkably improves T cell proliferation, persistence, and antitumor efficacy [17, 18]. This improvement can be attributed to the decreased number of lymphocytes; increased levels of homeostatic cytokines such as IL-7, IL-15 in the body; and reduced tumor cell activity [1, 6]. However, lymphodepletion has its own share of risks, including negative adverse effects [18, 23]. In contrast, in vivo generation does not necessitate such preconditioning because the targeted T cells need to be preserved for transfection. Furthermore, there is no competition between infused CAR-T cells and endogenous T cells for homeostatic cytokines [11, 16].

Limitations of in vivo CAR-T transfection

In vivo generation faces a significant limitation due to the lack of specificity of vectors and nanomaterials that could cause off-target transduction where viral vectors and nanomaterials recognize not only the target effector cell (i.e. T cells) but also other healthy cells, and worse of all, cancer cells [4]. A clinical study reported that CAR-resistant leukemia cells were detected due to off-target transduction of anti-CD19 CAR genes, resulting in their products binding to neighboring receptors and masking the CD19 epitope [5, 19]. Ultimately, this resulted in cancer relapse and patient death [19]. The exact implications of healthy cell expression of CAR genes remain uncertain. However, it could alter the way they react to external stimuli, induce abnormal behavior, and disrupt their usual cellular habits [4]. Although Stephan et al. have demonstrated that these CAR-expressing phagocytes do not expand, extensive research is warranted to gain greater knowledge on their expressions, potential life-threatening impacts, and ways to minimize off-target transduction [7].

The effectiveness of in vivo generation of CAR-T cells involves the delivery of CAR genes to the targeted T cells, which means that if the quality of the T cells from patients is poor, there might be no to little yield of CAR-T cells. Senescent T cells retain their ability to kill tumor cells but will not undergo expansion [20]. Patients' T cells may be of poor quality because of repeated exposure to tumors over an extended period, aging, or prior treatment [14].

In vivo expansion

There is a conundrum in CAR-T cell therapy. In a study involving diffuse large B-cell lymphoma (DLBCL) patients, 6%–24% were ineligible for CAR therapy due to inadequate ex vivo T cell expansion [27]. Furthermore, prolonged ex vivo culture and rapid expansion induce irreversible differentiation of T cells, linked to oxidative stress, impacting CAR-T cell functionality negatively. Currently, there are a few commercially available systems for ex vivo CAR-T expansion, including the GE WAVE system [26], G-Rex system gas-permeable bottles, and automated Prodigy machine [26].

Clinical information has illuminated the benefits of manufacturing less-differentiated memory T cell subpopulations on the efficacy and persistence of CAR-T therapy [37]. Stem cell-like memory T (TSCM) cells and central memory T (TCM) cells ensure prolonged CAR-T persistence, while effector memory T (TEM) cells drive immediate antitumor responses [48, 49]. In patients with lymphoma, naïve T cell (TN) or TSCM cells expressing CD8, CD45RA, and CCR7 correlated with better overall in vivo expansion of CAR-T cells [36]. Exhaustion markers negatively impact CAR-T cell function, highlighting the importance of using less-exhausted CAR-T cells. [36]. The use of TSCM and TCM cells devoid of exhaustion markers like PD-1, TIM3, or LAG3 has proven to be critical for the persistence and activity of CAR-T cells. It is vital to reduce the differentiation of T cells during manufacturing so that the ratio of various subpopulations can be optimized for enhanced therapeutic outcomes. In vivo expansion offers promise by circumventing prolonged ex vivo culture, thereby reducing terminal differentiation and enhancing CAR-T cell potency and persistence [28]. This has led to an influx of startups focusing on in vivo therapy and gaining high-value investments; this is illustrated in Table 7.1.

Advantages of in vivo CAR-T expansion

Ex vivo culture can span several weeks, and systemic delivery of CAR-T cells can lead to off-target effects that Macroporous Alginate Scaffold for T cell Engineering and Release "MASTER" seeks to address (Figure 7.2a). It is a biocompatible alginate scaffold with interconnected pores for optimal cell–virus interactions and efficient nutrient exchange for T cells and viral particles alike [4]. To activate T cells efficiently, DBCO-modified antibodies targeting CD3 and CD28 are immobilized on the scaffold's surface, eliminating the need for separate activation steps and saving invaluable time and resources. Furthermore, interleukin-2 (IL-2), a cytokine crucial for T cell proliferation, is encapsulated within the scaffold, and gradually released over five days to sustain the expansion of fully functional CAR-T cells [4]. MASTER is an integration of multiple ex vivo CAR-T cell manufacturing steps, including T cell activation, genetic modification, T cell proliferation, and release into a single in vivo scaffold. This technology does not require ex vivo culture, which would significantly reduce the overall cost of CAR-T cell therapy, making it a more viable treatment option. With localized in vivo expansion in the scaffolds, the risk of off-target effects is lowered, thus enhancing the safety and efficacy of CAR-T cell therapy.

Biomaterial scaffolds can also improve the rate of T cell expansion in vivo. Traditional T cell activation methods using microbeads or autologous monocyte-derived dendritic cells (moDCs) face issues of efficacy, cost, and variability. This ex vivo method often leads to limited T cell expansion rates and the generation of suboptimal T cell populations. Cheung et al. designed antigen-presenting cell-mimetic scaffolds (APC-ms) [31]

Table 7.1 Summary of startups developing in vivo CAR-T products

Company	Year started	Total raised [33]	In vivo products	Mode of operation [48]	Products in clinical testing	Product descriptions
UMOJA Biopharma	2019	$263M	UMOJA platform [46] • VivoVec, • Rapamycin-activated cytokine receptor (RACR) • TumorTag	Intended target cells • CD3+ CAR targets • CD19 • Prostate-specific membrane antigen (PSMA) • Carbonic anhydrase IX (CA IX) • Fibroblast activation protein (FAP) Route of administration • Intralymphatic (ILN)	ENLIGHTen • ENLIGHTen phase 1 trial assesses safety of universal CAR-T cells with UB-TT170 TumorTag [46].	Umoja's TumorTag tech uses bispecific molecules to mark tumor cells, enabling CAR-T cells to recognize and destroy them. UB-TT170 TumorTag targets folate receptors in osteosarcoma for improved treatment [46].
Capstan Therapeutics	2021	$165M	tLNP platform [42] • LNP delivery vehicle • Cell type-specific targeting binders • Disease-specific payloads	Intended target cells • CD2+ • CD8+ • CD5+ CAR targets • CD19 • CD20 • B-cell maturation antigen (BCMA) • FAP • Other undisclosed targets Route of administration • Intravenous (IV)	–	Capstan's safe lipid nanoparticle (LNP), a non-viral system, permits multiple in vivo doses. Its precise LNPs utilize cell-specific binders, such as antibodies, for efficient delivery of disease-specific payloads like CAR mRNA and therapeutic proteins [42].

Company	Year	Funding	Platform	Intended target cells	Bio's lead program	Description
Interius Biotherapeutics	2019	$76M	Bio's lead program [43] • I.NT2104	Intended target cells • CD3+ CAR targets • CD19 • CD20 • BCMA + • BCMA • Other undisclosed targets Route of administration • IV	Bio's lead program (INT2104) • Preclinical studies completed with high potential [47]	INT2104 is a lentiviral vector that is not self-replicating, carrying a gene for CAR20 targeting CD20. It is engineered to produce CAR-T and NK cells after one IV injection [43].
Enosoma	2021	$70M	Engenious platform [44] • Virus-like-particle (VLP) delivery system	Undisclosed	-	Engenious pioneers in vivo editing of hematopoietic cells through precise, one-time treatment. Their delivery system allows targeted and extensive payload transfer, setting the stage for future genomic medicines [44].
Vector Biopharma	2021	$30M	Shielded, retargeted adenovirus-based virus-like particle (SHREAD) platform [45]	Undisclosed	-	The SHREAD platform employs a non-viral particle and adapter proteins to precisely direct therapeutic DNA to specific biomarkers, enabling local, potent drug production while minimizing systemic toxicity for versatile gene-based treatments [45].

which combined lipid bilayers (SLBs) on high-aspect-ratio mesoporous silica micro-rods (MSRs) to form a three-dimensional (3D) scaffold (Figure 7.2b). The MSRs, with their substantial size and elevated aspect ratio, facilitated interactions with multiple T cells simultaneously for cell activation. This 3D scaffold enables fine-tuning of cue presentation and the microenvironment, thereby affording precise control over T cell expansion dynamics.

The pursuit of CAR-T therapy has revolutionized cancer treatment, yet it grapples with a significant challenge – the variable potency and durability of the antitumor response exhibited by autologous anti-CD19 CAR-T cells. This inherent variability arises from various factors, including T cell activation dynamics during production, patient health, and manufacturing intricacies [29]. APC-ms could be engineered to precisely regulate anti-CD3 and anti-CD28 stimulation density, enhancing T cell activation quality, and hence, personalizing the stimulation for each patient for optimal in vivo CAR-T expansion [29]. The personalized stimulation doses will be predicted based on a machine learning model of the T cell blood sample, APC-ms stimulation dose, and its resulting CAR-T cell products using data analysis techniques like classification modeling and random forest [29]. This advancement offers clinicians the ability to tailor the manufacturing process to the unique needs of each patient (Figure 7.2c).

Limitations of in vivo expansion

Although stimulating doses can be predicted to address the inherent variability in in vivo expansion, other sources can contribute to variability, making it a persistent issue. In a comprehensive exploration of in vivo expanded cells integrating either Bζ or 28ζ signaling domains, patients showed no differences upon examination, but the CAR structures from similar cellular sources exhibited remarkable diversity across patients [30]. Therefore, while a personalized stimulatory dose might enhance in vivo expansion, the type of signaling domain used can lead to variations in the final structure of the CAR-T product, increasing overall variability.

Clinical trial variability, stemming from factors like disparities in CAR-T cell manufacturing, and costimulatory domain selection, results in varied CAR structure combinations and behaviors in different patient cohorts [30]. Thus, it becomes more challenging to determine which signaling domain will trigger the best in vivo expansion results with maximal efficacy for each patient's specific CAR structure, which may further delay the therapy delivery. The delay occurs as subsequent strategies are taken to overcome these variabilities through standardization, personalization, comparative studies, and advanced CAR designs [30]. Standardizing manufacturing processes minimizes variability, while personalized approaches tailor CAR designs to individual immune profiles. Comparative studies could offer valuable insights to select the best CAR designs.

The absence of lymphodepletion before in vivo CAR-T cell manufacturing may complicate the creation of an optimal immunomodulatory environment for CAR-T proliferation and survival. Lymphodepletion is necessary for the recognition of target antigen through a single-chain variable fragment (scFv) binding domain, and without it, there will be a reduction in the T cell pool, an increase in immunosuppressive cell population, and a decrease in the availability of cytokines, therefore causing the fall in CAR-T proliferation and survival rates [38].

The biomaterial scaffolds described earlier utilize lentiviruses that are prone to activating innate and adaptive immune responses during in vivo expansion, hence possibly

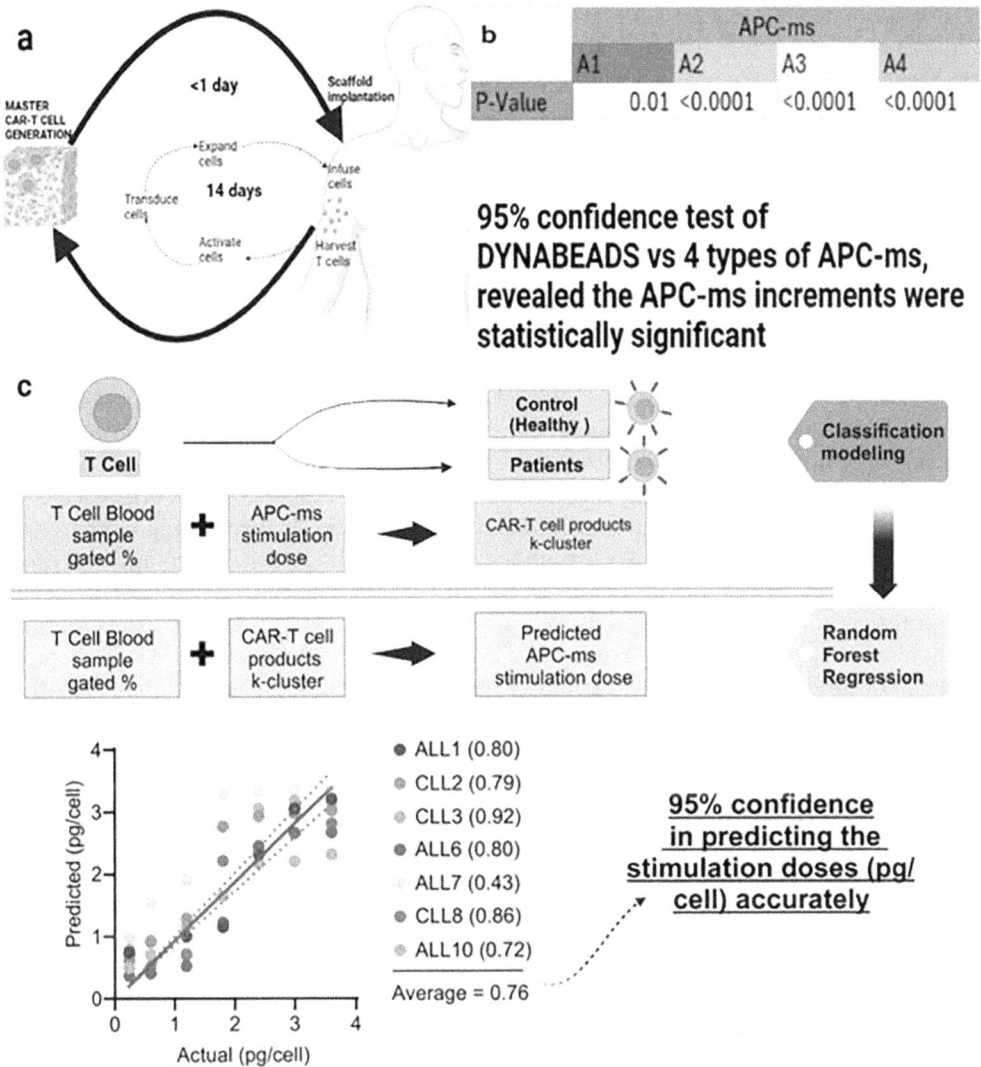

Figure 7.2 Technologies for in vivo CAR-T expansion (a) Schematic showing MASTER CAR-T Cell expansion method, which reduces the existing lengthy 14-day procedure to less than a single day, inspired by [4]. (b) Comparisons between antigen-presenting cell-mimetic scaffolds (APC-ms) and Dynabeads showing that APC-ms mimicked natural to achieve higher T cell expansion rates than Dynabeads [31]. (c) Schematic showing the process to predict personalized stimulation doses, inspired by [29]. A machine learning model utilizes data analysis methods such as classification modeling and random forest to forecast personalized stimulation doses. These predictions rely on information derived from the T cell blood sample, APC-ms stimulation dose, and the subsequent output of CAR-T cell products [29].

affecting the transduction efficacy of CAR-T cells [39]. LVs are known to trigger the inflammation and activation of innate immune sensors. The scaffolds used by MASTER contain retrovirus, which can also trigger immune responses that reduce in vivo CAR-T expansion.

The biomaterials required for fabricating the scaffolds must also be manufactured under current good manufacturing practices (cGMP) guidelines to ensure it is safe and strictly regulated. This adds a layer of complexity to CAR-T therapy as these materials may be costly to manufacture and cause immunogenic reactions [41]. A prime example would be stem cells where despite more than 20 years of research with material delivery, there is still to be an approved stem cell-based product (SCBP) by the US FDA [40].

Conclusion

In vivo generated CAR-T cells hold great promise for the future of CAR-T therapy, surpassing limitations of conventional methods, with evidence supporting their effectiveness being on par, if not greater. With numerous promising focus areas, in vivo transfection and expansion of CAR-T cells offer a pivotal opportunity to enhance and surpass existing limitations, thereby elevating its therapeutic efficacy. This success with in vivo CAR-T cells not only holds significant promise for enhancing current treatments but also ignites the prospect of extending similar methodologies to address diverse forms of cancer.

The use of biomaterials has advantages and disadvantages. Incorporating biomaterials into scaffolds enhances in vivo expanded CAR-T cell proliferation and survival. Nanoparticles made of biomaterials allow for the smooth delivery of CAR proteins for in vivo transfection. These biomaterials follow a stringent CGMP procedure to ensure their quality and efficacy. However, stringent CGMP regulation complicates CAR-T therapy, raising costs and lengthening the regulatory process. This reduces the availability of other products that could be beneficial to other patients and limits patients due to expensive treatment. Further streamlining of regulatory processes is vital for swift market access without compromising product quality.

Biomaterials in scaffolds and nanoparticles must be biocompatible to prevent immune reactions. Adverse effects like inflammation and autoimmune responses can hinder in vivo treatment and endanger patient health. Moving forward, we will require greater understanding of these cells and biomaterials interactions in vivo, to design scaffolds and nanoparticles with the appropriate physiochemical properties while reducing the immunogenicity of the final product.

Lymphodepletion reduces the efficacy of in vivo transfection, whereas it improves the efficacy of in vivo expansion. This process hurts the efficacy of in vivo transfection as it removes the target T cells in vivo, while promoting an optimal environment for in vivo expansion. Studies conducted often focus on in vivo transfection and expansion separately, causing a limited dataset for combined analysis. Bridging this gap is crucial for understanding how preconditioning of patients will affect overall treatment efficacy, necessitating additional efforts to link in vivo transfection and expansion effectively.

Patient variation is a pertinent issue that needs to be addressed in both in vivo expansion and transfection. Patient-specific varied T cell quality influences in vivo transfection, and these patient variations also lead to diversity in the CAR structure and antitumor effects exhibited by the in vivo expanded cells. Collectively, these affect the overall efficacy of the treatment. In certain scenarios, these patient-specific characteristics can prevent them from receiving in vivo treatment. While personalization of the treatment would ensure high efficacy, issues faced by the current personalized ex vivo treatment could recur. Through a greater understanding of the factors of variations, researchers can explore more ways to pinpoint the sources of these variations and identify treatment options to accommodate for these variations.

In vivo generation of CAR-T cells could be an effective methodology that overcomes the current limitations of ex vivo therapy. T cells can be manipulated in vivo to create CAR-T cells using an efficient strategy. The in vivo transfected cells can then be signaled and be provided with an optimal environment for efficient in vivo expansion. Overall, researchers' dedicated efforts show potential for revolutionizing CAR-T therapy, possibly becoming the conventional treatment, benefiting a broader audience.

References

[1] Cappell, K. M. & Kochenderfer, J. N. Long-term outcomes following car T cell therapy: What we know so far. *Nature Reviews Clinical Oncology* **20**, 359–371 (2023).

[2] Chen, Y.-J., Abila, B. & Mostafa Kamel, Y. CAR-T: What is next? *Cancers* **15**, 663 (2023).

[3] Agarwalla, P. et al. Bioinstructive implantable scaffolds for rapid *in vivo* manufacture and release of CAR-T cells. *Nature Biotechnology* **40**, 1250–1258 (2022).

[4] Niu, H., Zhao, P. & Sun, W. Biomaterials for chimeric antigen receptor T cell engineering. *Acta Biomaterialia* **166**, 1–13 (2023).

[5] Nawaz, W. et al. Aav-mediated *in vivo* car gene therapy for targeting human T-cell leukemia. *Blood Cancer Journal* **11**, 119 (2021).

[6] Mitra, A. et al. From bench to bedside: The history and progress of car T cell therapy. *Frontiers in Immunology* **14**, 1188049 (2023).

[7] Smith, T. T. et al. In situ programming of leukaemia-specific T cells using synthetic DNA nanocarriers. *Nature Nanotechnology* **12**, 813–820 (2017).

[8] Blache, U., Popp, G., Dünkel, A., Koehl, U. & Fricke, S. Potential solutions for manufacture of car T cells in cancer immunotherapy. *Nature Communications* **13**, (2022).

[9] Xin, T. et al. In-vivo induced car-T cell for the potential breakthrough to overcome the barriers of current car-T cell therapy. *Frontiers in Oncology* **12**, 809754 (2022).

[10] Pinto, I. S., Cordeiro, R. A. & Faneca, H. Polymer- and lipid-based gene delivery technology for car T cell therapy. *Journal of Controlled Release* **353**, 196–215 (2023).

[11] Parayath, N. N. & Stephan, M. T. In situ programming of car T cells. *Annual Review of Biomedical Engineering* **23**, 385–405 (2021).

[12] Pfeiffer, A. et al. In vivo generation of human CD19-CAR T cells results in B-cell depletion and signs of cytokine release syndrome. *EMBO Molecular Medicine* **10**, e9158 (2018).

[13] Neelapu, S. S. Managing the toxicities of Car T-Cell therapy. *Hematological Oncology* **37**, 48–52 (2019). doi:10.1002/hon.2595

[14] Nawaz, W. et al. Nanotechnology and immunoengineering: How nanotechnology can boost car-T therapy. *Acta Biomaterialia* **109**, 21–36 (2020).

[15] Dembski, S. et al. Establishing and testing a robot-based platform to enable the automated production of nanoparticles in a flexible and Modular Way. *Scientific Reports* **13**, 11440 (2023).

[16] Wall, D. A. & Krueger, J. Chimeric antigen receptor T cell therapy comes to clinical practice. *Current Oncology* **27**, 115–123 (2020).

[17] Davies, D. M. & Maher, J. Crosstown traffic: Lymphodepleting chemotherapy drives car T cells. *Cancer Cell* **39**, 138–140 (2021).

[18] Wang, A. X., Ong, X. J., D'Souza, C., Neeson, P. J. & Zhu, J. J. Combining chemotherapy with car-T cell therapy in treating solid tumors. *Frontiers in Immunology* **14**, 1140541 (2023).

[19] Wakao, R. & Fukaya-Shiba, A. In vivo car T cells and targeted gene delivery: A theme for the Pharmaceuticals and Medical Devices Agency Science Board to address. *Frontiers in Medicine* **10**, 1141880 (2023).

[20] Kasakovski, D., Xu, L. & Li, Y. T cell senescence and car-T cell exhaustion in hematological malignancies. *Journal of Hematology & Oncology* **11**, 91 (2018).

[21] López-Cantillo, G., Urueña, C., Camacho, B. A. & Ramírez-Segura, C. Car-T cell performance: How to improve their persistence? *Frontiers in Immunology* **13**, 878209 (2022).

[22] Khan, I., Saeed, K. & Khan, I. Nanoparticles: Properties, applications and toxicities. *Arabian Journal of Chemistry* **12**, 908–931 (2019).

[23] Depil, S., Duchateau, P., Grupp, S. A., Mufti, G. & Poirot, L. 'Off-the-shelf' allogeneic car T cells: Development and challenges. *Nature Reviews Drug Discovery* **19**, 185–199 (2020).

[24] Huckaby, J. T. et al. Bispecific Binder redirected lentiviral vector enables *in vivo* engineering of CAR-T cells. *Journal for Immunotherapy of Cancer* **9**, (2021).

[25] Walters, P. Engineering angles: Countering the viral vector shortage. *Pharma Manufacturing* (2020). Available at: https://www.pharmamanufacturing.com/sector/large-molecule/article/11298750/engineering-angles-countering-the-viral-vector-shortage. (Accessed: 17th November 2023)

[26] Wang, X. & Rivière, I. Clinical manufacturing of car T cells: Foundation of a promising therapy. *Molecular Therapy - Oncolytics* **3**, 16015 (2016).

[27] Herda, S. et al. Long-term in vitro expansion ensures increased yield of central memory T cells as perspective for manufacturing challenges. *International Journal of Cancer* **148**, 3097–3110 (2021).

[28] Ito, Y. & Kagoya, Y. Epigenetic engineering for optimal chimeric antigen receptor T cell therapy. *Cancer Science* **113**, 3664–3671 (2022).

[29] Zhang, D. K. et al. Enhancing car-T cell functionality in a patient-specific manner. *Nature Communications* **14**, (2023).

[30] Cheng, Z. et al. *In vivo* expansion and antitumor activity of coinfused CD28- and 4-1bb-engineered car-T cells in patients with B cell leukemia. *Molecular Therapy* **26**, 976–985 (2018).

[31] Ruella, M. et al. Induction of resistance to chimeric antigen receptor T cell therapy by transduction of a single leukemic B cell. *Nature Medicine* 24, 1499–1503 (2018).

[32] Cheung, A. S., Zhang, D. K., Koshy, S. T. & Mooney, D. J. Scaffolds that mimic antigen-presenting cells enable *ex vivo* expansion of primary T cells. *Nature Biotechnology* **36**, 160–169 (2018).

[33] Fidler, B. 'in vivo' cell therapy: Expanding beyond car-T. BioPharma Dive (2022). Available at: https://www.biopharmadive.com/news/in-vivo-cell-therapy-car-t-biotech-startups/634287/. (Accessed: 17th November 2023)

[34] Berger, C et al.: Adoptive transfer of effector CD8$^+$ T cells derived from central memory cells establishes persistent T cell memory in primates. *The Journal of Clinical Investigation* **118**, 294–305 (2008).

[35] Gattinoni, L et al.: A human memory T cell subset with stem cell-like properties. *Nature Medicine* **17**, 1290–1297 (2011).

[36] Xu, Y et al.: Closely related T-memory stem cells correlate with *in vivo* expansion of CAR. CD19-T cells and are preserved by IL-7 and IL-15. *Blood* **123**, 3750–3759 (2014).

[37] Turtle, C. J. et al. CD19 car–T cells of defined CD4+:CD8+ composition in adult B cell ALL patients. *Journal of Clinical Investigation* **126**, 2123–2138 (2016).

[38] Liu, Q., Liu, Z., Wan, R. & Huang, W. Clinical strategies for enhancing the efficacy of car T-cell therapy for hematological malignancies. *Cancers* **14**, 4452 (2022).

[39] Follenzi, A., Santambrogio, L. & Annoni, A. Immune responses to lentiviral vectors. *Current Gene Therapy* **7**, 306–315 (2007).

[40] George, B. Regulations and guidelines governing stem cell based products: Clinical considerations. *Perspectives in Clinical Research* **2**, 94 (2011).

[41] Giancola, R., Bonfini, T., & Iacone, A. Cell therapy: cGMP facilities and manufacturing. *Muscles, Ligaments and Tendons Journal*, 2(3), 243–247 (2012).

[42] Developing and delivering precise *in vivo* cell engineering to patients. *Capstan Therapeutics* (2023). Available at: https://www.capstantx.com/#js-refuse-cookies. (Accessed: 2nd December 2023)

[43] Andorko, J. I., Gill, S. & Johnson, P. Targeted *in vivo* generation of car T and NK cells utilizing an … *Interius Biotherapeutics* Available at: https://interiusbio.com/wp-content/uploads/2023/06/Interius-Bio-Data-Poster-June-2023-1.pdf. (Accessed: 2nd December 2023)

[44] The engeniousTM platform. *Ensoma* Available at: https://ensoma.com/engenious-platform/. (Accessed: 2nd December 2023)

[45] Home. *Vector Biopharma* (2023). Available at: https://www.vectorbiopharma.com/#technology. (Accessed: 2nd December 2023)

[46] Policano, J. Umoja Biopharma Announces Activation of first enlighten phase 1 trial site in study of UB-TT170, a tumortag that targets folate receptors to mark tumors for clearance by

car T cells. *Umoja Biopharma* (2022). Available at: https://www.umoja-biopharma.com/news/umoja-biopharma-announces-activation-of-first-enlighten-phase-1-trial-site-in-study-of-ub-tt170-a-tumortag-that-targets-folate-receptors-to-mark-tumors-for-clearan/. (Accessed: 2nd December 2023)

[47] Interius biotherapeutics highlights strong preclinical data supporting *in vivo* chimeric antigen receptor (CAR) vector evaluation in clinic. *Cision PR newswire* (2023). Available at: https://www.prnewswire.com/news-releases/interius-biotherapeutics-highlights-strong-preclinical-data-supporting-in-vivo-chimeric-antigen-receptor-car-vector-evaluation-in-clinic-301856091.html. (Accessed: 2nd December 2023)

[48] Blair, W. In Vivo CAR -T Competitive Landscape. *William Blair Equity Research* (2023). Available at: https://media.licdn.com/dms/image/D4E22AQGhlYjx_HHYSQ/feedshare-shrink_2048_1536/0/1704714461272?e=1707955200&v=beta&t=nrMBRkmneugFWPwFoTEztDYiMKUACcRFnEAckUIAm4s. (Accessed: 11th January 2024)

Index

Pages in *italics* refer to figures and pages in **bold** refer to tables.

For Product Safety Concerns and Information please contact our EU
representative GPSR@taylorandfrancis.com
Taylor & Francis Verlag GmbH, Kaufingerstraße 24, 80331 München, Germany